The Magic of the Truffle

The Magic of the Truffle

The Favorite Recipes of Christian Etienne

Master Cuisinier of France
Provençal Cook

ici la
PRESS

Published by Ici La Press
694 Main St. South
Woodbury, CT 06798
www.icilapress.com

Printed in Spain by Imago
LCCN 2001090557
ISBN 1-931605-00-9

10 9 8 7 6 5 4 3 2 1

To all those who helped me
in the trade,
to all those who taught me
to become what I always had in my head:
to be a cook.

Warmest and special thanks to
Philippe Parc, the best tradesman of France,
world champion of desserts,
for his generous collaboration
on the chapter on desserts.

Recipes and intoductory text collected
by Robert Ledrole.

Contents

Preface

Christian Etienne's food, his restaurant, his personality—one could perhaps even say his soul—are permeated with the smells and tastes of his beloved Provence. When I met him for the first time, I knew immediately that this man could not have come from anywhere else in the world.

Christian has written beautifully about other of his favorite regional ingredients such as the tomato, but, as all chefs know, the truffle is one of the most beguiling ingredients in the world. One cannot remain indifferent to it—people either love the taste and smell of this aromatic tuber or simply loathe it. I love truffles, cook with them whenever the opportunity presents itself, and can even become practically poetical when talking about them, but Christian takes this love to new heights. We are all fortunate that he decided to transcribe this love onto paper.

Christian's deep passion for cooking, that of the truffle in particular, comes across in this wonderful book, full of his knowledge and of the sunshine of Provence. He shares his recipes and imparts his experience with generosity of spirit and humor. I devoured many of his dishes, unfortunately vicariously, while reading this cookbook cover to cover. I am sure that you too will come under the spell—both of the truffle and of Christian Etienne himself.

Daniel Boulud
June 2001

Some General Points on Truffles and Their Use

Truffles are hard to find and quite expensive, so you have to be a bit savvy buying them at the local market. From time to time gullible shoppers are sometimes taken in by shady purveyors, who offer truffles of poor quality at the same high prices as the good ones.

Stories abound of clever merchants deceiving buyers. Some of these stories are no doubt true; others have been fabricated. One hears of truffles stuffed with lead shot to increase their weight, of truffles pieced together from various parts to hide defects, of white truffles, of less value, that have been rubbed with charcoal to disguise their true color. . .

These tricks are made possible by the fact that truffles remain mysterious things that live underground and are bought and sold on the quiet. It's as though they need to be shrouded in mystery to maintain their value.

Before you shop for truffles, do your homework so you know exactly what you're looking for, and above all, avoid acting like a haughty Parisian when you're making your purchase. If you've managed, despite all, to acquire some truffles, here are a few principles to know before using them.

Storing Truffles

Once a mature truffle has been dug from the ground, it doesn't keep long. It will lose its aroma and qualities after ten days.

To keep a truffle fresh for two or three days, wrap it in paper towel, put it in an airtight plastic container, and set it in the refrigerator. Do not clean a truffle until you are ready to use it. If you put eggs with a truffle to impart its flavor to the eggs, be sure their shells are perfectly dry and wipe any moisture that forms on them daily. Avoid condensation and humidity, which can cause truffles to rot.

If you aren't going to use a truffle within a few days, the best thing to do is to freeze it. Wash it, brush it, dry it with care, then wrap it tightly in aluminum foil and place it in the freezer.

Types of Truffles

The most attractive truffles aren't necessarily the best. Some are crooked or bumpy and some appear to have an elongated neck. The explanation for the lumps and bumps can be found in the soil. As truffles grow, they often encounter stones or other obstacles, which they must grow around.

What separates good truffles from bad is not so much their appearance as their maturity and their aroma. Knowledgeable truffle lovers want to examine the flesh of a truffle before they buy it and will ask that a small cut be made for that purpose. Their aim is to evaluate the firmness, color, and structure of the flesh, all of which vary according to the species of truffle.

A truffle's aroma, while a good indicator of quality, cannot be relied upon alone. Truffles are living things, and the strength of their perfume can change from one day to another. A truffle that has no scent at all may, the following day, release the most glorious of aromas.

Here are the truffles you're most likely to find.

Tuber melanosporum

The queen of truffles is black with a warty skin and finely marbled flesh. It is harvested between November 15 and the end of February. The best time to harvest it is usually January, but one never knows with this diabolic tuber.

Tuber uncinatum, Burgundy Truffle

Black with large, pointed warts in a form of a diamond. The flesh is closer to chocolate than it is to black. The aroma is less powerful than that of melanosporum but quite delicate.

Tuber aestivum, Summer Truffle

This truffle is available in May and June and sometimes in July if the weather isn't too dry. It is clearly inferior to the preceding types, but can be used to make good salads.

Tuber brumale, Musk Truffle

A horror? It has a disgusting odor, and its taste will spoil any dish.

No one buys it on purpose, but it bears a strong resemblance to melanosporum, both inside and out, and is harvested at the same time and in the same areas. The two species are easily confused. Buyer beware.

How to Prepare Truffles

Colette, who was an avid gourmet, called truffles "black princesses." She felt strongly, as did Georges Sand (also a fine gastronome), that the truffle, a noble food, should only be used with products of very high quality to avoid bad marriages.

Since you don't eat truffles every day (would you even if you could?), you owe it to yourself, even for something as simple as scrambled eggs, to make the meal like a little special event, fussing over every detail. The meal will appear to be the better for it.

When you buy or otherwise obtain a fresh truffle, it is often covered with soil. You must brush it to remove the soil, which is often embedded in the rough surface.

Then you rinse the truffle in warm water and dry it thoroughly.

To prepare it, you can cut it into thin strips with a knife, a mandolin (if you have one for making potato chips), or with a special tool made for this use that allows you to vary the thickness of the slices.

Truffles can also be cut in other ways according to needs of the recipe:

In matchsticks [*en bâtonnets*]**:** first cut the truffle into somewhat thick strips, then into matchsticks.

In a julienne: matchsticks that are sliced again into the thinnest strips possible.

In cubes or dice: of varying size. They are cut from very thick matchsticks.

In a *brunoise*: tiny dice. To arrive at these small pieces, mince a julienne.

Try, no matter how you prepare the truffles, to make regular cuts—it makes the dish look more attractive.

It's also important to understand that ⅓ oz. (10g) of a fine *brunoise* provides as much flavor as 1⅓ oz. (40g) of truffles cut into large cubes.

Whenever possible, cut the truffle just before you use it. A truffle reduced to a *brunoise* loses its aroma very rapidly—better it impart its aroma to the other ingredients than to the atmosphere.

You will likely note that there are almost no recipes in this book that require prolonged cooking. Truffles do not improve when cooked for long periods.

Their savory and aromatic qualities are released as soon as they are warmed through, not cooked. High temperatures do nothing for them at all. Colette was right. Truffles are princesses that deserve to be pampered and that go perfectly with seasonal foods. The exception may be wild game, which requires high temperatures (roasts) or long cooking times (stews, casseroles, etc.) and whose strong, gamy taste fights with that of the truffles.

Vegetables work nicely with truffles. Leeks, celery, mâche salad, and endives know how to surround them with affection. Truffles show the same sort of affection for white meats, fish, and shellfish.

Truffle cuisine is born of these relationships, which permit the cook to create happy marriages of flavors. This festive cuisine is sometimes a little rich, but it is not the truffles that are the source of the calories.

Meals at which truffles are present are joyful occasions that foster a special bond among the guests.

Now it's your turn to create this joy and share it with the fortunate people who sit at your table.

Soups

Cream of celery

Crème de céleri

Ingredients for 6 servings
2.2 lbs. (1kg) leeks, white part only, washed – 1 cup olive oil
2.2 lbs. (1kg) celery hearts – 2 cups (½ liter) white stock
2 cups (½ liter) water – 6 slices sandwich bread with the crust removed –
4.5 oz. (115g) butter – 7 oz. (200g) truffles, julienned – salt – pepper

Cut the whites of the leeks into very thin slices and, in a medium saucepan, sauté them gently in olive oil.

Then cut the celery into small dice, add these to the leeks, and cover with the stock and an equal amount of water.

Let the soup cook. During that time cut the bread into small cubes and fry them in butter. The resulting croutons will give the creamy soup a bit of crunch.

When the celery is thoroughly cooked and soft, pour the soup into the food mill (or use a blender, but a blender incorporates too much air into the vegetables, changing their flavor).

Then incorporate 2 oz. (60 g) of the butter with a whisk, add the julienned truffles, which will give the soup an exquisite flavor.

Adjust the seasoning to taste and serve over the croutons. This celery soup will warm the spirits of your guests!

A word from the chef:

To make a good cream of celery, you really need a food mill. Potatoes can be substituted for the celery.

Cream of pumpkin

Crème de courge

Ingredients for 6 to 8 servings
9 oz. (250g) onions and 7 oz. (200g) leeks (white part only),
washed and drained – ¼ cup olive oil – 2.2 lbs. (1kg) pumpkin flesh only
4 cups (1 liter) light cream – salt – pepper –
7 oz. (200g) truffles, julienned

Mince the onions and leeks, place them in a saucepan, and sauté in olive oil add a bit of salt. (The onions and leeks must be chopped very fine because, in this recipe, the soup will not be passed through a food mill.)

Then cut the pumpkin flesh into large cubes and add them to the saucepan. Why cut the pumpkin into large cubes? Otherwise they will quite simply melt.

After you put the pumpkin in the saucepan, fill the pan half way with the cream and cook over low heat. The pumpkin will bit by bit turn to mush.

Stir occasionally to help the pumpkin soften. Then season to taste and continue to stir until the soup has a creamy consistency.

The color is splendid! You can add more cream and allow the soup to thicken slowly on the stove.

Don't put it in the food mill; this soup is meant to have a little texture.

Add the truffles just before serving. Truffles should always be used nearly raw. That's how they impart the best flavor to the other ingredients.

Variations

You can, if you like, serve the soup over croutons, which are prepared by frying small cubes of sandwich bread in butter.
Even better, choose a small pumpkin, such as Jack-Be-Little, remove the stem, and scoop out the strings and seeds. Fill the raw pumpkin with soup, add salt and pepper and a touch of garlic, and cook in a medium oven for a couple hours to obtain the same result. The soup is served in the pumpkin, which acts as a tureen.

A word from the chef:

If you can, use dishes in complementary colors: the soup will be orange with thin strips of truffles.
This soup should not be put through a food mill.

Cream of mussels with truffles

Crème de moules aux truffes

Ingredients for 4 servings
4.4 lbs. (2kg) cultivated mussels – 2 cups (½ liter) chicken broth
1 cup heavy cream – 3½–4 oz. (100–120g) truffles
freshly ground pepper

The first step, of course, is to wash the mussels and carefully remove the beards.

Drain them well and then steam them in a pot until they open.

Take the mussels out of their shells and strain the cooking liquid. Reserve a few mussels for decoration and purée the rest in a blender.

Thin the purée with the mussel cooking liquid, or if the liquid is too salty, use chicken broth. Chicken broth is better than fish stock to preserve the mussel taste. Next mix in the heavy cream that has been whipped briskly as for whipped cream.

Julienne the truffles very fine and add them, and freshly ground pepper, to the hot soup.

Decorate the soup with the reserved mussels and with four thin slices of truffle, whose color will contrast with that of the soup.

Cabbage soup with pork sausage and truffles

Soupe au chou à la saucisse de couenne truffée

Ingredients for 4 or 6 servings
1 green cabbage, medium size – 3 medium carrots
3 cloves peeled garlic – 2 medium onions studded with cloves
2.2 lbs. (1kg) large pork sausage – salt – pepper – croutons made from
pain de campagne – 9 oz. (225g) truffles, julienned – fine olive oil

Here's a soup that will recall the kind that was once made in the hearth. It has a bit of fat in it to help keep you warm in winter.

Begin by preparing the cabbage, removing the outer large green leaves, washing the head, and cutting it in 4 wedges.

Then put the cabbage in a large pot, so it has plenty of room. Add water, to cover carrots, garlic, the clove-studded onions and the sausage. Bring to a very light boil and skim the soup.

The cabbage must cook down very slowly. This isn't a clear soup; it should be more like a fondue. The ingredients need to blend their flavors in the broth. Adjust the seasoning to taste.

While the soup is cooking, prepare the croutons by cutting the bread into cubes and frying them in butter, or toasting them in the oven until they are golden.

Place croutons in each dish and on top of them the julienned truffles.

Remove the sausage from the soup, cut it into small cubes, add them to each dish.

Pour the broth, which should be very hot, over the croutons, truffles, and sausage. Sprinkle with a good dose of pepper and a drizzle of first quality olive oil, this soup shouldn't be bad, not bad at all!

A word from the chef:

You can find excellent pork sausages at a butcher shop that specializes in pork.
Go easy when seasoning the soup; the sausage already contains a fair amount of salt and pepper.
If you have some big, old, country bowls, this would be the time to bring them out.

Oyster soup

Soupe d'huîtres

Ingredients for 6 servings
18 large oysters − 3 oz. (80g) butter − ¾ cup (100g) flour
4 cups (1 liter) spicy chicken broth reduced to 2 cups (½ liter)
juice of ½ lemon − grated nutmeg − freshly ground pepper
4 oz. (115g) truffles

Open the oysters and reserve and strain their juice.

Next, make a velouté sauce by first preparing a white roux. Melt the butter, sprinkle it with flour, and stir briefly. For the roux to remain white, the flour must not cook too long nor too quickly. Allow the roux to cool before combining it with the chicken broth; if the roux isn't cool the velouté becomes lumpy.

Whisk the roux and broth together and put the resulting velouté on the stove over low heat for half an hour.

Strain the velouté through a fine chinois and adjust its consistency by either adding broth or putting the velouté back on the stove to reduce further. You want the velouté to have the consistency of a light cream.

Do not add salt. Instead, season with pepper and nutmeg.

Thin the velouté with the juice of the oysters, add the lemon juice, and put it back on the stove.

Drain the oysters again carefully for the second time for their liquid. Place the oysters in the bottom of each dish and arrange them with the truffles. Finally, pour the hot soup over the oysters. The boiling liquid will plump the oysters without cooking them too much and also pick up the powerful aroma of the truffles.

A word from the chef:
*There are four kinds of roux: white, blond, brown,
and completely ruined!*

Oxtail soup with celery and truffles

Soupe de queue de boeuf au céleri et aux truffes

Ingredients for 4 servings
3 lbs. (1½kg) 1 large oxtail – 1 carrot, peeled – 1 cup (100g) celery hearts
10½ oz. (300g) truffles, diced – 1 large piece puff pastry dough
salt – freshly ground pepper

Prepare the oxtail and make a broth with it.

Cook it like a pot-au-feu with the usual aromatic mixture of vegetables and herbs—onion studded with a few cloves, mixed herbs, black peppercorns, and soup vegetables, which you will drain, chop, and mix with the other ingredients.

While the broth is cooking, cut the carrot and the celery into small dice. Sauté these vegetables in olive oil for 5 minutes.

Preheat the oven to 400°F.

Place the diced truffles, the carrots and celery, the seasoned broth, and meat in a tureen.

Cover the soup with a thin layer of puff pastry dough.

Place the tureen in the oven.

When the puff pasty is cooked, serve.

The moment the pastry is broken, the strong aroma of this soup will fill the room.

Frog legs in a chicken velouté

Velouté de blancs de volailles aux grenouilles

Ingredients for 6 servings
3 oz. (80g) butter – ¾ cup (80g) flour – 4 cups (1 liter) chicken stock
18 frog legs – 2 cups (½ liter) milk – 2 cups (½ liter) crème fraîche – salt
freshly ground pepper – ½ lemon, juice only – nutmeg
3½ oz. (100g) truffles, julienned

Begin by making a velouté. To do that, prepare a white roux with the butter and flour. Add the chicken stock and allow the resulting velouté to cook over low heat for at least an hour.

A long, slow cooking time will completely remove the flour taste.

Meanwhile, poach the frog legs in milk. One boiling is enough. Drain the frog legs and pat them dry. Then remove the bones with care and put the meat in the serving tureen along with the julienned truffles.

When the velouté has finished cooking, add the crème fraîche, whisking firmly but gently.

To finish the velouté, put it back on the stove; it should be smooth and white. Adjust the seasoning and add the juice of half a lemon.

Grate just a bit of nutmeg into the velouté and stir. Pour over the frog legs and truffles.

The sudden heat will release the heady perfume of the truffles.

Serve immediately in large soup dishes.

Croutons and green garnishes are superfluous.

A word from the chef:

Here's a recipe that will make your best friend girlfriend die of jealousy.
If you have English friends over, you'll be a huge success!
They won't believe it's frog legs.

Appetizers

White sausages

Boudins blancs

Ingredients for 12 to 15 servings
1¼ lbs. (600g) onions, peeled – 2 chicken breasts, preferably from young,
free-range chickens – 4 slices sandwich bread, crust removed – salt
pepper – 3 oz. (80g) truffles – white chicken stock – butter

Sauté the onions until they are transparent; you don't want them to take on any color.

Chop the chicken breasts into a small dice.

Cook them with the onions until they become just a bit firm.

Dip the bread in the milk. Wring it out and mix it with the chicken and onions. Chop the mixture very fine.

Add the truffles *brunoise* to the mixture.

Screw the funnel onto your meat grinder and carefully fit the end of a length of sausage casings over the opening. The casing must not tear. Tie the other end with heavy sewing thread.

Then push the mixture into the casing. Twist the sausage and tie it closed every 2 in. or so to make small sausage links.

Poach the links for 20 to 30 minutes in the white chicken stock just at the boiling point. Remove the links and drain them.

Then puncture them with a fork and fry them in butter. Serve them golden brown.

A word from the chef:

There is no boudin sausage without onions.
Ask your butcher for sheep casings.
Arrange to borrow a meat grinder with a funnel.

Brandade of salt cod

Brandade de morue

Ingredients for 6 to 8 servings
1 lb. 2 oz. (500g) salt cod − 9 oz. (250g) fresh cod
4 thick slices ginger − zest of orange and lemon − 1 cup milk
1 cup cream − 1 cup olive oil − 1 ½ to 2 oz. (40–60g) truffles, chopped
croutons made from pain de campagne − 1 clove garlic
truffle strips for garnish

Soak to desalt, then drain the salt cod. Remove the bones with a pair of tweezers then flake with a fork.

Remove the skin of the fresh cod. This can be done easily by holding the flesh and pulling on the skin. Flake the fresh cod as well.

Mix fresh and salt cod together with the ginger and the orange and lemon zest.

Heat the mixture in the milk.

In another saucepan, reduce the cream.

When the milk boils, wait for the second boil and take the saucepan off the heat.

Remove the ginger and zest and blend the fish with an electric mixer stick blender or by hand. Now put the fish in a large mortar or cooking pot and begin stirring constantly while slowly adding the reduced cream followed by the olive oil. Your arm will ache, but it will be worth it!

Finally, stir in the chopped truffles. Because the *brandade* is still warm, the flavor of the truffles will permeate the entire mixture.

To serve, spread the mix on croutons that have been rubbed with a clove of garlic and lightly toasted.

You can decorate and garnish the brandade with thin strips of truffles.

A word from the chef:
To make this dish, begin the day before by soaking the salt cod in a large tub to remove the salt. Change the water several times.

Chicken heart brochettes

Brochettes de coeurs de volailles

Ingredients for 4 servings
12 chicken hearts (3 per brochette) – 4 oz. (115g) truffles
olive oil – 1 cup Port sauce with truffles

Cut the hearts in two down the middle. Make sure there is no more blood and discard any pieces of arteries or veins.

Slice and *brunoise* the truffles very fine.

Marinate the hearts with the truffles in a first quality olive oil about an hour.

Push the hearts onto small skewers that you will sauté rather than braise.

Sauté quickly at very high heat, then dip immediately into a Port sauce.

Game & green vegetable packets

Caillettes de gibier

Ingredients

cooked wild game (the meat will count for half of the combined weight of the 2 vegetables. For example: if you use 4 oz. (115g) Swiss chard and 8 oz. (230g) spinach you should use 6 oz. (345g) of meat) – ⅓ (by weight) Swiss chard leaves – ⅔ (by weight) spinach – caul ⅙–⅓ oz.(5–10g) truffles per packet – 1 cup broth – a few sage leaves

Remove the bones carefully. Blanch the vegetables.

Cut the truffles into very small cubes while the caul soaks in clear water.

Then combine the meat and vegetables and chop them up. Add the truffles and mix them in so that everything is blended evenly.

Next, fashion very small packets. Contrary to normal practice, try not to make them much larger than a quail's egg.

Wrap the packets with caul, set them in a covered cooking dish with broth, and braise them in the oven for one hour. Then sprinkle with sage leaves. This dish makes a hot hors-d'oeuvre with quite a strong flavor. It will perk you up!

A word from the chef:

You can mix different kinds of game, both birds and animals. The game will be cooked. This is a dish that's easy to prepare and very practical, because it allows you to use the leftover scraps of meat.

Carpaccio with truffles

Carpaccio de tête de veau aux truffes

Ingredients for 6 to 8 servings
½ calf's head (with tongue) – bouquet garni, thyme, bay leaf
1 stalk of celery – 1 carrot – 1 onion studded with 6 cloves
truffles, sliced and pounded – truffle vinagrette

Carefully soak the head in vinegar water for several hours. Then simmer it with a mixture of vegetables and herbs—thyme, bay leaf, celery, a carrot, the onion studded with cloves—over low heat for 2 hours.

Once the head is cooked, put it on an oven rack that has been placed on a sheet pan covered with a clean cloth.

Even out the thickness of the head by cutting pieces from the thickest part and placing them on the thinnest part. Remove the tongue and even out its thickness in the same fashion, cutting the pointed end off and placing it next to the base of the tongue.

Cover the tongue with truffles. Then put the tongue in the middle of the head and roll the head into a tight cylinder.

The thickness of the head must be uniform in order to make an even cylinder around the tongue (in the shape of a roast).

Wrap in a sheet of parchment paper or plastic and place in the refrigerator for one day to give the truffles time to flavor the meat. The next day, cut into very fine slices with a rotary meat slicer, or with a sharp thin slicing knife, or ask your butcher to cut the carpaccio for you.

Arrange the slices in the center of the plates and serve with a truffle vinaigrette. Makes a cold appetizer that is out of this world.

Chicken liver flan

Flan de foies de volailles

Ingredients for 6 to 8 servings
7 oz. (200g) chicken livers − 6 eggs − 4 cups (1 liter) cream − salt
freshly ground pepper − nutmeg freshly grated − 2 shallots
1 clove garlic − salt − 3½–4 oz. (100–120g) truffles
truffled Port sauce

Blend all the ingredients except the truffles in a food processor or blender. Be careful not to overheat the livers or introduce too much air into them. Pulse several times.

Season with 2 tsp. (10g) salt, 1 tsp. (3g) pepper, and a little grated nutmeg. Then pour the liquid through a fine strainer.

Cut the truffles into juliennes.

Lightly butter the interior of cylindrical or baba molds.

Fill them with the liver mix ¾ full and cook in a water bath at 300°F for two hours.

Let the flan cool and then put the molds in the refrigerator for 3 days to allow the flavors to blend and become more refined. Reheat slowly, unmold, and serve with a truffled Port sauce.

A word from the chef:

Blending is the same as mixing; a blender or immersion stick is a mixer for a professional chef.
For this recipe, you need small cylinder or baba molds.
You also need 3 days for the flan to rest in the refrigerator. It will be all the better for the wait!

John Dory in a pastry crust

Pâté en croûte de Saint-Pierre

Ingredients for 1 pâté en croûte
For the dough:
5½ cups (800g) all-purpose flour − 7 oz. (200g) lard
1 egg yolk − 1 cup scant water − 1 tbsp. scant sea salt
For the filling:
¾ lb. (335g) John Dory fillets − 2 oz. (60g) large truffles
1 lemon, juice only − salt − pepper − olive oil − 1 egg, beaten
cold butter − 3 cups gelatin
mâche − truffle vinaigrette

On a pastry board, arrange the flour to form a well in the center. Place the lard in the circle and work it into the flour with your hands. Add the egg yolk and continue to knead. Add water as needed. At the last moment, add the salt. Finally, cut the dough in half: Flatten each ball of dough with the palms of your hands, pressing in all directions. Roll back into a ball and repeat.

Place the dough in the refrigerator under a cloth.

The next day, clean the John Dory fillets and remove the skins. Season with salt and pepper and sauté in hot olive oil until firm. thirty seconds on each side, no more.

Reserve the fillets as soon as they are cooked and allow them to cool.

Roll out the dough on a table dusted with flour and use it to line a pâté en croûte mold (one that opens on the sides).

Slice the truffles into thin slices. Fill the mold with a layer of John Dory fillet followed by a layer of truffles and continue to alternate back and forth. Between each layer, pour a bit of lemon juice. Finish with a layer of fish.

Close the dough over the pâté and decorate the dough with leftover pieces of dough cut into fish shapes. Then make a chimney in the top of the pâté by pushing a rolled up business card through the top dough.

Brush the dough with beaten egg and cook in an oven preheated to 300°F (150°C) for 45 minutes. Remove from the oven and allow to cool.

Prepare the gelatin and pour it into the pâté while it is still warm.

A good trick: If the dough splits open, fill the crack with cold butter to keep the gelatin from running out.

Remove the butter when you take the pâté out of the refrigerator, by that time the gelatin will have solidified. Serve with some mâche salad leaves napped with a truffle vinaigrette.

A word from the chef:
This is the way to make a pastry pâté. It's made with lard.
Commercial flours vary in gluten content. So add water to the dough,
but make certain that it stays on the dry side.
Prepare the dough the day before you need it.

Savory chicken liver parfait

Parfait de foies de volailles

Ingredients for 1 terrine
10 oz. (300g) chicken livers – 2 cups (½ liter) white wine
3½ oz. (100g) truffles – 1 cup, 3 tbsp. (300g) butter, softened
20 slices pain de compagne, toasted

Wash the chicken livers. Blanch them in the white wine; one boil is enough. Drain well.

Slice the truffles into juliennes. Run the livers through the food processor. Then put the minced livers into a frying pan, not to cook them but to keep them warm over low heat.

Fold the butter into the liver.

Add the truffles and stir thoroughly. Place the mixture in a buttered or plastic wrap–lined terrine and put it into the refrigerator for 24 hours. Serve on toasts.

A word from the chef:

Choose a butter of excellent quality; it plays a key role in this recipe. Don't indulge too heavily in this parfait. It is very, very high in calories!

Toasts with truffle spread

Toasts au petit ragoût de truffes

Ingredients for 2 to 4 servings
1 medium onion − 2 oz. (60g) butter − 1 oz. (30g) flour
1 cup good white wine − 4–6 slices of good bread − 1 oz. (30g) truffles
1 large garlic clove − salt − pepper

Chop the onion fine and sauté lightly in half of the butter.

When the onion is transparent, add the flour, and cook for a few moments.

Add the white wine diluted 2 to 1 with water. Allow the mixture to reduce until it is almost dry.

Toast the bread.

Grate the truffle into the sauce and stir a bit to encourage the flavors to blend. Lightly rub the toast with garlic. Then butter it and top it with the truffle spread.

A word from the chef:
You can cut the bread in the form of a heart or another shape before toasting it.

Eggs

Scrambled eggs with truffles

Brouillade de truffes

Ingredients for each serving
3 eggs – salt – freshly ground pepper – 1 oz. (30g) truffles, sliced
1 oz. (30g) olive oil or butter – 1 tsp. crème fraîche

Break the eggs. Beat the eggs lightly and season with salt and pepper.

Slice half of the truffles. Mix in the sliced truffles and allow the eggs to sit for half an hour.

Prepare the water bath. Cook the eggs as slowly as possible over water bath in a little olive oil or butter. This is an exercise in patience. Stir the eggs constantly.

Add a bit of crème fraîche to give the eggs a creamy quality, nothing more. When the eggs are nearly done, slice the rest of the truffles on top, turn them into the eggs to blend them together thoroughly, and serve at once.

A word from the chef:
Scrambled eggs demand your absolute attention!
When they are ready, don't make them wait. Ideally
they should be cooked in a ceramic skillet over a water bath
or on an electric hot plate. The secret to scrambling eggs
is to cook them slowly.

Creamed eggs with parmesan and truffles

Crème aux œufs et au parmesan

Ingredients for 2 servings
6 eggs – seasoned rich broth or meat – 2 oz. (60g) parmesan, grated
juice – freshly ground pepper – 3 oz. (80g) truffles, cut into tiny dice

Separate the egg whites and yolks into two bowls.

Beat the egg whites longer than you do the yolks. Mix the whites and yolks together.

In a cast-iron frying pan, reduce the broth almost to a glaze. Allow it to cool.

Add the eggs, then the grated parmesan, to the broth and season with pepper (with the parmesan, you'll need little or no salt).

Put the pan back on the stove.

Whisk lightly, add the truffles, and continue to stir over low heat to give the mixture a smooth consistency.

A word from the chef:

Truffles go nicely with parmesan, as long as the parmesan is of the highest quality and is grated just before use.

Gratin of hard-boiled eggs

Gratin d'œufs durs

Ingredients for 6 servings
2–3 oz. (60–80g) truffles – 6 hard-boiled eggs
½ cup thick crème fraîche – salt – pepper

Chop the truffles roughly.

Peel the eggs and cut them in two along their length.

Remove the yolks. Mash the yolks, mix them with a little crème fraîche and the truffles, and season to taste.

Refill the eggs with this mixture, which will make a very pleasing little mound.

Coat the bottom of a baking dish with crème fraîche and place the eggs inside, stuffing side up. Brown in the hot oven for a few short minutes.

Truffle omelet

Omelette aux truffes

Ingredients for each serving
3 eggs – 1½ oz. (40g) truffles, sliced – salt – pepper
olive oil or butter – frisée lettuce salad with garlic

Prepare the omelet by beating the eggs (but not too much).

Mix the eggs with half of the sliced truffles. Salt and pepper to taste. Allow the flavors to blend for half an hour.

Cook the omelet on high heat in a nonstick pan with a little olive oil or butter.

Shake the pan over the fire to keep the omelet from sticking.

When it's time to roll the omelet, sprinkle the remaining truffles over its entire surface. Roll the omelet. Serve and eat!

A salad of frisée lettuce delicately flavored with garlic is almost imperative with this omelet.

A word from the chef:

This omelet is served rolled, with the center a bit runny.

Omelet soufflé

Omelette soufflée

Ingredients for each serving
2 eggs − 2 tsp. sugar − 2¾ oz. (80g) truffles, sliced − 2 oz. (60g) butter

Separate the yolks from the egg whites.

Whip the whites into stiff peaks by beating hard with half of the sugar.

Add the other half of the sugar, with the truffles, to the yolks.

Delicately mix the beaten egg whites with the yolks. Cook in a frying pan with a noisette brown butter. Roll and serve sliced.

A word from the chef:

Undertake this sweet appetizer and follow it with fish in a cream sauce accompanied by small glazed turnips. Serve with a somewhat syrupy white wine.
Champagne would also go well with this meal.

Shellfish

Petits gris snails in a spinach ragout

Escargots petits gris en ragoût d'herbes

Ingredients for 4 servings
2 lbs. (1kg) spinach − 3 oz. (80g) butter − salt
freshly ground pepper − nutmeg, freshly grated − 48 petits gris snails
(12 per serving) − 2 cups (½ liter) heavy cream
1 loaf pain de campagne − 1 clove garlic − truffles

Wash the spinach, remove the coarse stems and blanch. Drain and chop, drain and reduce in brown butter.

Season with salt, pepper, and a bit of nutmeg.

Rinse and drain canned snails or remove snails from their shells that you've gathered from your garden. Reduce the heavy cream by half on the stove for a few minutes.

Add the spinach to the cream and cook a little more over very low heat. Next add the snails and mix thoroughly.

Toast large slices of bread rubbed with garlic, without pressing too hard. The flavor of the garlic must blend with the flavors of the other ingredients, not overpower them.

Spread the spinach and snails over the toast and cover with strips of truffles.

Place in the 450°F oven for 2 minutes (only) and serve.

A word from the chef:

You can find canned snails of excellent quality that are ready to use. If you prefer, you can gather your own, deprive them of food for a while, feed them grass or flour, wash them, and finally add them to a court bouillon with fennel and other herbs.

Oysters gratin with truffle sabayon

Huîtres gratinées au sabayon de truffes

Ingredients for each serving
3½ oz. (100g) spinach – melted butter – salt – freshly ground pepper
nutmeg, freshly grated – 6 oysters – 1 egg yolk
2 oz. (60g) butter – ¾ oz. (20g) truffles, julienned

Wash the spinach and remove the coarse stems. Then wilt it in a sauté pan with some melted butter.

Add salt, pepper, and a bit of nutmeg, which goes nicely with the spinach.

When the spinach is cooked, drain it under a weighted board to squeeze out the water.

Meanwhile, open the oysters, reserving the liquor, and set them aside. Strain and reserve the liquor and wash the shells.

In the bottom of each shell, make a soft bed of spinach and place the oyster on top, arranging it attractively.

Now prepare the sabayon with truffles. Begin by beating the egg yolk with the oyster liquor. Cook the mixture over a water bath stirring constantly to make an emulsion.

To determine whether the sabayon is cooked, dip a finger into it. If it's hot to the touch, it's done. Stop cooking and add butter cut into small pieces whisking constantly.

Place the julienned truffle on each oyster in its shell and cover with the sabayon.

Place in a hot oven for 5 to 8 minutes, until the sabayon is golden.

A word from the chef:
To put the oysters in the oven, set them on an oven tray that has been covered with a layer of salt. The salt will keep the oysters from tipping.

Roasted sea scallops

Saint-Jacques rôties

Ingredients for each serving
3 large scallops − 1 oz. (30g) truffles − 1 tbsp. olive oil
2 tsp. balsamic vinegar − 1 cup salad (mâche or endive)

Begin by cutting the scallops after having prepared and cleaned them for cooking.

To cut them, place them flat on a cutting board and cut them horizontally as though you are going to make a sandwich, without cutting all the way through. Do not separate into two halves.

Then slice the truffles into somewhat thick slices and put 1 slice between each scallop. The truffle should appear sandwiched by the cut scallop.

Next, sear the scallops in a sauté pan in which you've preheated olive oil. Take note: you must cook the scallops quickly; they must never be overcooked. The frying pan should be smoking hot, so that the scallops cook instantly. A light brown caramel will form on their surface, giving a slight hazelnut flavor.

When you bite into the scallop and truffle, the two flavors will blend to create a glorious sensation . . .

Deglaze the frying pan with a little balsamic vinegar. Serve the scallops with the mâche salad or endives dressed with the warm pan vinaigrette.

Scallop salad

Salade de Saint-Jacques

Ingredients for 2 servings
6 large scallops − 3½ oz. (100g) truffles, cut into 24 slices
coarse sea salt − freshly ground pepper
juice of ½ lemon − 4 tsp. olive oil, first quality

Slice the scallops horizontally very thin.

Arrange them in a circle on 2 plates, interspersing them with the slices of truffles.

Season with salt and pepper. Heat the plates lightly to give a little warmth to the ingredients. Then dress with a vinaigrette made from lemon juice and olive oil.

A word from the chef:

For a romantic dinner. choose two attractive plates that go well with the colors of the salad.

Littlenecks and truffle toasts

Toasts aux tellines et aux truffes

Ingredients for 6 servings

4½ lb. (2kg) small littleneck or Manila clams – 12 thick pieces sandwich bread with the crust removed – 4 oz. (115g) truffles – butter – pepper

Rinse the clams in a tub of clear water. Change the water several times and discard all the shells that don't look good. Drain carefully.

Then cook the clams just as they are, without butter or oil, in a large sauté pan in a single layer. Don't pile them on top of each other. If you have more than will fit easily in the pan, cook in batches.

Stir constantly until the shells have all opened, taking each from the pan as its shell opens and removing the meat from the shell. Save the liquid from the clams as they open.

Scoop out the center of the bread slices before toasting. Chop the truffles and roll them in your hands with softened butter. Strain the juice from the clams. Thin the butter with some of this juice. Cover the bread with the clams and spread the truffle butter over them.

Place under the broiler for 5 minutes before serving.

A word from the chef:
You need a very large sauté pan to cook the clams.
Be careful not to overcook them.

Note: Tellines are a very small type of clam only found in the sea water of the Carmagne area of Provence. Substitute small littleneck or Manila clams.

Crab gratin

Gratin de Tourteau

Ingredients for each serving
1 Dungeness crab, about 1 lb. (450g)
For the court bouillon:
1 carrot − 1 stalk of celery − 1 slice lemon − 1 medium onion
8 cups (2 liters) waters − dash of pepper
For the Mornay sauce:
2 cups (½ liter) milk − 1 egg yolk − 2 oz. (50g) butter
3 oz. (80g) truffles − salt − pepper

1 oz. (30g) grated parmesan − truffle strips

Cook the crabs in a court bouillon. Drain them thoroughly.

Then open them and remove the meat from the body cavity and set the legs aside. Wash, rinse, and dry the shells.

Prepare a Mornay sauce by first making a béchamel sauce with a white roux and milk. Then add the egg yolk, tempering it with the sauce, and whisk in butter. Season to taste.

Preheat the oven 450°F.

Grate the parmesan and shred the truffles and cut in fine *brunoise* dice.

Mix the crabmeat with the sauce and add the truffles. Fill the shell cavity with the mixture. Sprinkle grated parmesan on top and brown briefly in a very hot oven, with the tray set fairly high in the oven.

Garnish with strips or sticks of raw truffles.

A word from the chef:

You can also present this gratin in small ramekins if you want to serve it as a small appetizer. In that case, one crab will serve 2 or 3 people. This dish can also be made with spider or spiny crabs.

Note: Tourteau is a large crab found in Europe; Dungeness is a good substitute in the U.S.

Prawn gratin

Gratin de queues de langoustines

Ingredients for each serving
3 prawns − I cup white chicken stock − I small carrot
I small stalk of celery − 2–3 oz. (60–80g) butter − salt
freshly ground pepper − 2 egg yolks − I cup white wine
truffles, julienne

Remove the prawns from the shells. Crush the tails and reserve.

Then prepare a broth with the white chicken stock and the crushed prawn shells. Reduce the broth by one quarter.

Meanwhile, cut the carrot and celery into tiny slivers and cook them gently in butter.

In a sauté pan, cook the prawn tails in a little clarified butter.

When the stock is reduced, strain it carefully.

Then mix the sautéed vegetables, drained of their cooking butter, with the prawn tails and adjust the seasoning with salt and pepper.

Beat the egg yolks over a water bath into a sabayon.

Combine all the ingredients. Place in a gratin mold. Make sure that the juice is not too reduced. Fold in the truffles julienne. Place in the oven for a few minutes to color. The eggs in the sabayon will thicken the sauce.

A word from the chef:

Note: Prawns can not be mediocre.
Use white chicken stock in preference to fish stock. The flavor of a fish stock will be too strong.

Lobster in cream

Homard à la crème

Ingredients for 2 servings
Two 1 to 1½ lb. (450–675g) lobsters − olive oil − pepper, freshly ground
1 cup cognac − 2 cups (½ liter) crème fraîche − 2 tbsp. butter, softened
3 oz. (80g) truffles, julienned

Split the heads of the lobsters in two, in parallel with the body, by striking them with a cleaver. The lobsters won't suffer a bit.

Remove the tails by pulling away from the heads. Reserve the coral, if there is any.

Remove the pocket or stomach of gravel that the lobsters may have picked up along the way. Remove the claws and flatten them on the cutting board by hitting them (not too hard) with the flat side of the cleaver.

Put olive oil in a large sauce pot. When the oil is almost smoking hot, drop in the lobsters (tails, heads, and claws).

Stir until the shells are red. Add lots of freshly ground pepper.

Then douse the lobsters with cognac and set them on fire. Move the pot so that all the parts are flambéed.

Lower the heat and add the crème fraîche. Cook for 10 minutes.

Take out the tails and the claws, remove the meat, and put to one side, keeping it warm.

Break the shells up into large pieces and put them back in the pot to cook with the crème fraîche. The sauce should turn pink.

Next, pour the sauce through a fine-mesh conical strainer (china cap) into another saucepan. Work the coral with a little softened butter and push it through a fine sieve over the sauce, pressing with a spoon.

Put the sauce through the conical strainer again. Slice the lobster tails thin and place on two plates. Spoon the sauce around. Add the julienned truffles. Voila! It's ready to serve!

A word from the chef:

You're going to flambé the lobsters. The origin of this practice goes back a long way. The flesh of animals was flambéed with alcohol to remove the last hairs adhering to the skin.
In olden times people flambéed the flesh they offered to their gods.
The use of the coral as a thickening agent is comparable to the use of blood in a civet. It's important that the coral not be overcooked.

Prawns in zucchini robe

Langoustines en robe de courgettes

Ingredients for each serving
1 small zucchini – 3 prawns – 3 oz. (80g) truffles
salt – freshly ground pepper – olive oil – truffle vinaigrette

Wash the zucchini. Very carefully cut long slices of skin from the zucchini. Try to make the slices as fine as possible.

Reserve the uncut zucchini flesh. Remove the prawns from their shells. Keep the last segment intact (so you can grab the tail).

Cut the truffles in fine strips.

Lightly salt and pepper the prawn tails.

Wrap the tails first with truffles then with the translucent strips of zucchini skins, which will protect the truffles during cooking.

This is a little tricky to do, but worth the trouble. Heat up the steamer.

Grate the flesh of the zucchini, add salt and pepper, and drain a bit. Then form balls with the zucchini flesh—one per guest—and fry them in olive oil at very high temperature.

Cook the tails of the prawns in the steamer for 5 or 6 minutes. In the center of each plate, place the fried zucchini. Arrange 3 prawn tails radiating outward. Finally, sprinkle with a truffle-flavored vinaigrette.

A word from the chef:
You'll need a couscoussier or a steamer with a lid.

Note: Very large prawns or spiny lobster tails can be substituted for langoustines.

Spider crab salad

Salade d'araignée de mer

Ingredients for 2 or 3 servings
I spider crab – 2 carrots, sliced – I small onion, sliced
I stalk of celery, sliced – I thick slice lemon – peppercorns
3 oz. (80g) truffles, cut into bâtonnets
2 tbsp. truffle vinaigrette – 2 cups mâche lettuce

Prepare the spider crab. Cook it in a court bouillon that includes the sliced carrots and celery, the whole onion, a thick slice of lemon, plus a few peppercorns to give the broth a little zip.

Drain carefully and keep the broth for future use.

Remove the meat from the body cavity and legs of the crab. Mix with the truffle bâtonnets.

Place the mixture in the center of each plate, surround with mâche salad and dress with the vinaigrette.

A word from the chef:

*This can just as well be made with regular crabs or lump crabmeat.
In Provence, no one has ever heard of a spider crab; to get one, you have to ask for an esquinade.
You can make a fish broth by breaking up the crab shell and adding it to the court bouillon. Reduce by one quarter, strain through a fine mesh strainer and freeze.
It's always good to have broth on hand!*

Fish

Sole puff pastry turnovers with truffles

Chaussons aux paupiettes de filets de soles truffés

Ingredients for 4 servings

1 carrot, medium – 1 shallot – 1 stalk of celery – 1 onion, medium
3 oz. (80g) truffles – butter – 4 sole fillets – salt
freshly ground pepper – 1 or 2 puff pastry sheets – 1 egg yolk

Cut the vegetables into small *brunoise* and mix together.

Cut the truffle into small bâtonnets.

Sauté the vegetables lightly in butter. Salt and pepper to taste.

Preheat the oven to 350°F (180°C).

Roll out the puff pastry with a rolling pin.

Flatten and dry the sole fillets.

Place mixed vegetables and truffles on top of each fillet. Roll the fillets and lay them on the puff pastry, and form it into a turnover.

Seal the edges of the turnover with egg yolk.

Prick the top with a fork, then glaze with a pastry brush dipped in egg.

Place the turnover in the oven on a tray dusted with flour for 15 minutes.

A word from the chef:

There are several different ways of making sole fillet turnovers. Here are two:
The one above blends three flavors—those of the fish, vegetables, and truffles. The resulting taste is complex as the truffle will be the dominant flavor, while the shallot and carrot will provide a slightly sweet touch that is accentuated by the celery.
The second and ostensibly simpler recipe follows. This combines only the flavors of the fish and the truffles, each of which intensifies the other.
The choice is yours.

Sole turnovers

Chaussons de sole

Ingredients for 6 servings
6 sole fillets – salt – freshly ground pepper – 6 oz. (180g) truffles, chopped – ½ lb. (225g) puff pastry or ½ pkg. phyllo sheets – 1 egg yolk truffle-flavored Port sauce

This dish is easy to make. All you need to do is cover the sole fillets with the chopped truffles, then salt and pepper and roll them.

Make turnovers from the puff pastry or phyllo and sole, glaze them lightly with egg yolk and put them in the oven just as they are! The savory flavor will come from the juice of the truffles and the sole inside the turnovers.

Bake between 10 to 15 minutes in preheated 375° oven.

Serve the turnovers in the center of a large, brightly colored plate and surround them with a truffle-flavored Port sauce.

And if you don't say they're delicious, I'll eat my hat.

A word from the chef:
You can use very small truffles for this recipe.

Monkfish liver terrine

Foie de lotte

Ingredients for 1 terrine
2¼ lbs. (1kg) Monkfish liver (ask your fish market to put aside
a nice one for you) – 2 cups (½ liter) milk – salt
freshly ground pepper – a few drops of Port or cognac
5 oz. (150g) truffles, cut into thin slices – capers

Soak the liver in milk for 2 hours.

Blot dry, then remove the skin and the nerves without damaging the flesh.

Season precisely according to the following formula: 2 generous tsp. (12g) salt and just shy of 1 tsp. (4g) pepper per kilo (2¼ lbs.).

Put a few drops of Port or cognac into a terrine.

Set the liver inside, inserting a few slices of truffles between the lobes.

Then cook for an hour at 250°F (120°C).

Allow to cool.

To serve, cut in thin slices to keep truffles and liver together. Capers should provide a pleasant contrast with the liver.

A word from the chef:

In northern France, the monkfish or anglerfish is called lotte, *which may also be spelled* lote.
In the south, the fish is known as baudroie.
This recipe is based on the one for goose or duck livers.

Red mullet livers

Foies de rougets

Ingredients for 1 small terrine
1 shallot, finely diced − olive oil − 7 oz. (200g) red mullet livers
milk to cover − salt − freshly ground pepper − 1 baguette
2 oz. (60g) truffles

Sauté the shallot in a little olive oil until it becomes transparent. Do not let it turn brown.

Soak the livers in milk for 2 full hours. Cut the truffles into thin strips and set aside.

Drain the livers well in a colander and clean them by pulling on the nerve that runs down the middle.

Chop the livers with a knife. Mix the chopped livers with the shallots and heat over low heat without cooking. Season with salt and pepper.

Cut the baguette in two and toast it lightly.

Preheat the oven to 375°–400°.

Spread the pieces of liver thickly over the baguette and add the strips of truffles.

Place in the oven until heated through.

Serve immediately.

A word from the chef:
The same dish can be made with chicken livers.

Note: Rouget is a highly regarded fish found in the Mediterrean, you could use regular mullet or chicken livers.

Monkfish tail studded with truffles

Gigotin de lotte

Ingredients for 4 servings
3 oz. (80g) truffle − One 3½-4 lbs. (1,6–1,8kg) monkfish tail, bone in
salt − freshly ground pepper − 2 cups (300g) flour − 2 egg yolks, beaten
3 cups very fine white bread crumbs − 1 cup clarified butter
sautéed potatoes with truffles

Cut the truffle into small bâtonnets.

Press the truffle sticks with a small sharp knife into the flesh of the fish. Season with salt and pepper.

Dust with flour and dip in beaten egg yolks, then in the bread crumbs.

Sprinkle with clarified butter and place in a hot oven for 15 to 20 minutes, depending on thickness.

Turn the fish over halfway through cooking and baste with more butter to give the fish a golden brown color. Serve with sautéed sliced potatoes with truffles.

The fish and potatoes work well together.

A word from the chef:
The coating of bread crumbs will prevent the truffle sticks from pushing out of the fish.
If you want to impress your friends, tell them that you have caught the monkfish.

Salmon tartare

Tartare de saumon

Ingredients for each serving
3½ oz. (100g) raw fresh salmon fillet − 1½ oz. (40g) truffles
3 whole oysters − 1 lemon − 1 bunch chives, cut into 2-in. (5-cm) lengths

Chop the salmon very carefully but quite finely with a knife.

Chop the truffles less fine. Open and drain the oysters, sprinkle with lemon juice, and chop them as well.

Make a nest of the salmon. Put the truffles and oysters in the middle and finish the presentation with chives.

Dim the lights and eat without another moment's delay.

The Truffle
on Its Own

Truffle purses

Aumônières de truffes

Ingredients for 20 purses
10½ oz. (300g) finely ground fresh veal – butter – 1½ oz. (50g) truffles,
roughly chopped – 2 artichoke bottoms, cooked and rubbed with lemon
1 tbsp. armagnac – salt – freshly ground pepper – 1 lb. (450g) pasta
dough or 1 pkg. wonton skins – chives (to tie the purses together)

Lightly brown the veal meat in butter.

Chop the truffles and the artichoke bottoms in big pieces.

Mix the meat, truffles, and artichokes in a terrine. Spinkle with the armagnac, salt lightly, and pepper with a bit of a heavy hand.

Cover and set in refrigerator for an hour.

Prepare the pasta dough and roll it as finely as possible on a cutting board dusted with flour.

Cut out 3 in. (7.5 cm) circles of dough.

Put a bit of the veal stuffing in the center of each circle.

Then fold each one closed in the form of a beggar's purse.

Tie the crimped part of the purse with one or two chive stems. Steam in a steamer or a couscoussier for about 20 minutes.

Truffle fritters

Beignets de truffes

Ingredients
4½ oz. (125g) flour – 1 egg – fine salt – 1 clove garlic
1 oz. (30g) truffles per person – oil for frying – gray salt flavored with
iodine from the Guérande region

Rub your bowl with garlic before you make your fritter batter. When this batter is cooked, it must take on a garlic aroma without having a garlic taste—yes, that's right!

Then mix the flour, egg, and a pinch of fine salt in the bowl.

Thin with some water if needed but don't add anything else.

Allow the batter to sit while you cut the truffles into thick strips. Check the consistency of the batter; it must be light.

Heat the oil to 375°, then dip the truffles separately in the batter. Drain them. Let them drip a second. Then fry them until the coating begins to turn golden.

Remove the fritters with a wire skimmer and serve them as soon as they have drained. Put 3 or 4 large grains of sea salt crystals on each fritter before taking a bite. The salt provides a delightful crunch.

A word from the chef:
These fritters make a nice appetizer when you're having
a few friends over.
They're nice, a little rich, yes, but oh so nice.

Truffle in rich chicken stock

La truffe au bouillon de poule

Ingredients for each serving
1 oz. (30g) truffles − 1 casserole dish, with cover and a hole, per serving
excellent rich chicken stock, clarified and reduced
salt − freshly ground pepper − 1 tbsp. each flour and water

Put a truffle in a small casserole. The truffle should just fit inside.

Add stock until the casserole is three quarters full. Adjust the seasoning to taste.

Cover the casseroles and seal them hermetically by wrapping them with a length of dough made from flour and water.

Put the casseroles in an oven preheated to 375°F for 10 minutes. Serve immediately.

A word from the chef:

Make sure the truffles are all more or less the same size to prevent jealous squabbles among your guests!

Truffles with Port and cream in covered ramekin

Cassolette de truffes fraîches au porto et à la crème

Ingredients for each serving
1–2 oz. (30–60g) fresh truffles − salt − pepper
2 tsp. heavy cream − 2 tsp. Port
1¾ oz. (50g) puff pastry dough − 1 egg yolk, beaten

Cut the truffles into quarters. Salt and pepper them, and place them in 3 in. (7,5 cm) ovenproof ramekin.

Combine the cream and Port and reduce by half.

Pour into the ramekin, filling halfway up the truffles. Cover with a thin layer of puff pastry dough.

Prick the dough with a fork and glaze it with beaten egg yolk. Cook in the oven at 400°F. Serve in the ramekin on a napkin on a plate.

A word from the chef:
To be savored with respect.

Truffled eggs in their shells

Julienne de truffes fraîches en coques d'œufs

Ingredients for 4 servings

4 eggs – ½ cup heavy cream – salt – freshly ground pepper
2 oz. (60g) truffles – 1 tbsp. each pine nuts and blanched almonds, finely chopped – 2 oz. (60g) butter for toast – sippets made of the highest quality bread or cut from toasted sandwich bread

Begin by carefully cutting the tops of the eggs then separating the whites and the yolks into separate bowls.

Beat the yolks with the cream, add salt and pepper.

Then beat the whites into stiff peaks.

Slice the truffles into a fine julienne, reserving 4 slices, then divide into 4 equal portions, and put 1 portion in each shell. Add the mixture of yolks and cream to each shell and heat for 2 minutes in a hot oven 400°F.

Remove the shells without turning off the oven and carefully put the beaten egg whites on top of each egg. Sprinkle with the pine nuts and almonds and put the eggs back in the oven for 2 more minutes, no more.

Then slip a reserved slice of truffle into the egg whites. Serve with buttered toasted bread triangles.

A word from the chef:

Cooked in the egg shells, which must be reserved after you remove the tops and empty them of their contents. You'll need egg cups. This is an agreeable recipe that will amaze your friends.

Truffle and mushroom ragout

Ragoût de truffes fraîches et champignons de Paris

Ingredients for 2 servings
1¾ oz. (50g) fresh truffles − 6 large white mushroom caps
1 lemon, juice only − 1 cup Port wine − salt
freshly ground pepper − 1 egg yolk

Dice the truffles and the mushroom caps. Keep separate. Sprinkle
the mushrooms with lemon juice and cook in a saucepan with the Port.
Salt and pepper to taste.
Reduce the cooking juice by half.
Add the diced truffles. Stir just three or four times.
Remove from the fire and thicken with an egg yolk.
Do not cook any further (because of the egg). Serve immediately.

A word from the chef:
This dish can be served as an accompaniment to a red meat.

Truffle salad

Salade de truffes

Ingredients for each serving
2½ tbsp. balsamic vinegar – 4 tbsp. olive oil, first quality
salt – freshly ground pepper – 3½ oz. (100g) truffles per serving
4 cups mâche or frisée – fine Guérande sea salt

Prepare a somewhat spicy vinaigrette.

Cut the truffles into thin slices of equal size and place with the mâche leaves on individual plates or in a salad bowl. Pour the vinaigrette over.

Salt the salad (after it has been served) with fine sea salt.

For a more rustic salad dice the truffles and mix them with frisée lettuce in a bowl to experience a delicious contrast in flavor and color.

A word from the chef:

You can make splendid presentations with slices of truffles overlapping each other by a third and laid out in a spiral on the plate. Or, for a more natural look, you can serve this salad in a Provençal Aubagne honey colored glazed salad bowl.

Peasant truffle toasts

Tartine de truffes à la paysanne

Ingredients for 4 servings
1 qt. (1 liter) red Châteauneuf-du-Pape wine − 2 shallots, minced
2 oz. (60g) butter − 1 qt. (1 liter) rich stock − 7–10 oz. (200–300g) truffles
4 large slices pain de campagne, toasted

Begin by reducing the wine by three fourths. That will take some time during which you can sauté the shallots in butter until they turn transparent.

Add the stock to the shallots and reduce by half.

Combine the two reductions.

Ther resulting sauce should look like a syrup.

Chop the truffles roughly and heat them in a saucepan.

As soon as the truffles are hot, add the sauce. Allow it to come to a boil, then remove immediately from heat.

Spread over lightly toasted slices of peasant bread and serve with a bottle of the same Châteauneuf-du-Pape used to make the sauce.

The wine and toast go nicely together.

A word from the chef:

Peasant flavors, but only for peasants who are quite well off!
In this recipe, you can use very small truffles.

Truffles with parmesan

Truffes au parmesan

Ingredients for 4 servings
4 fairly large truffles − 1 cup extra virgin olive oil − salt
freshly ground pepper − 4 oz. (120g) parmesan, first quality

Cut the truffles into thin slices.

Marinate them in olive oil for 4 hours. Then preheat a 350°F oven.

Pour a little of the olive oil from the marinade into a small gratin dish.

Then add a layer of sliced truffles. Add salt and pepper. Grate just enough parmesan to cover the truffles.

Then add a few drops of oil. Repeat this procedure, finishing with a layer of parmesan.

Place in the preheated oven for 15 minutes.

Serve as a hot hors-d'oeuvre with a bottle of asti, or a red Barbera that is lightly spumante, or a bubbly rosé.

A word from the chef:
Select an aged parmesan of very high quality.
Select a fairly small gratin dish, 3 in. (7,5 cm) oval, this must be at least
4 layers. Keep the leftover oil to season a watercress salad in which
you add a scant amount of a good vinegar.

Truffles in bread crust

Truffes en croûte de pain

Ingredients for each serving
1 truffle weighing 1¾–2 oz. (50–60kg) – 1 piece pain de campagne,
a bit stale – a few drops of red Port wine
1 tsp. crème fraîche – a pinch of salt – freshly ground pepper

You're about to prepare a very simple but quite extraordinary hors-d'oeuvre that will please your nose as much as it will your taste buds.

The first thing to do after you've gotten the ingredients together and cleaned the truffles is to cut the pain de campagne into very thick slices—at least ¾ in. (2 cm).

Then preheat the oven to 450°F; it must be very hot.

Cut circles in the bread that will fit as caps for the ceramic egg cups. The trick is to make the bread caps fit the cups as closely as possible.

Then put a few drops of Port, in each egg cup with a truffle, and a teaspoon of crème fraîche. Season with salt and pepper and fit the egg cups with the bread.

Place in the oven and leave the thermostat set high.

When the bread begins to brown, serve quickly without burning yourself. Break through the toasted bread. It's fabulous!

A word from the chef:
You'll need small ceramic egg cups that are oven safe.

Truffle packets

Truffes en papillote

Ingredients for each serving
1 oz. (30g) truffles — freshly ground pepper
2 thin slices slab bacon

Brush clean the truffles.

Dry them thoroughly. Pepper them with gusto.

Wrap them with slices of bacon, put them in a foil pouch packet, crimp closed on all the edges, set in an oven preheated to 400°F.

Depending on the thickness of the truffles, leave them in the oven for 10 to 15 minutes.

The strips of bacon will make you think you're eating a real piece of paradise. As for the truffles . . .

If you have a fireplace, cook the papillotes over the coals. They won't be any better, but they'll go well with the ambience.

A word from the chef:
This is one of the simplest ways of preparing truffles, which are particularly appropriate in a country setting where wild game is being served.

The raw truffle

La truffe crue

Ingredients for each serving
the best pain de campagne – the finest of churned butters
1 oz. (30g) truffles, cut into thin slices – fine Guérande sea salt

Place slices of bread in the oven just long enough to toast them a bit.

Butter the bread.

On top of the butter, place a reasonable thickness of truffle slices, then a few grains of salt.

Taste and savor!

Simple as this recipe is, there is a variation: Leave out the salt and spread your slices of bread with the best salted butter you can find.

A word from the chef:

*The shame of the cook! There is no cooking in this recipe!
It's a snack and a way to taste truffles before moving on to more serious recipes.*

Stuffed Things

Stuffed zucchini

Courgettes farcies

Ingredients for 4 to 6 servings
4 medium zucchini – 1 medium onion – ¼ cup olive oil
1 or 2 cloves garlic, minced – bread crumbs – 1 sprig (very little is
needed) – rosemary or a few sage leaves, finely chopped
truffle, finely diced – 2 tsp. aged balsamic vinegar

Wash the zucchinis and cut them into 2-inch-thick rounds.

Carefully remove the flesh with a spoon and reserve.

Blanch the zucchini rounds in salted boiling water. Leave them only for a second boil.

Fill a salad bowl with ice water and shock the zucchini rounds in it. Drain in a colander then on a cloth.

Sauté the onion in olive oil, with the garlic and the reserved zucchini flesh. Cook long enough to evaporate the juice from the zucchini flesh. If necessary, add a pinch of bread crumbs to absorb the liquid.

After cooking, add the rosemary and the truffle.

Mix everything together thoroughly and stuff the zucchini rounds.

Heat the stuffed zucchini in a 400° oven and serve with a few drops of aged balsamic vinegar.

A word from the chef:
Once it is blanched, zucchini becomes fragile.
Handle it with care. Chilling the zucchini quickly fixes the chlorophyll,
preserving the attractive green color of this vegetable.

Stuffed squid

Encornets farcis

Ingredients for each serving
2 tbsp. cooked rice pilaf per person − ½ tsp. truffle oil − 1 or 2 squids
(depending on size) − ¼ cup white wine − ¾ oz. (20g) truffles
salt − freshly ground pepper − 1 oz. (30g) butter
1 cup each of mâche salad and frisée − 1 tbsp. lemon dressing

Make a rice pilaf flavored with truffle oil.

Clean the squid bodies without cutting them open. Clean their tentacles and cut them into small cubes.

Cook the diced tentacles in a dry white wine.

Meanwhile, cut the truffles into a fine dice.

Mix the truffles with the rice pilaf.

Next add the diced squid tentacles to the stuffing. Adjust seasoning to taste with salt and pepper.

Stuff the squid and close them with a small skewer or a toothpick.

Brown them in a sauté pan with butter. Carefully turn them on all sides.

Serve in an attractive presentation of frisée mixed with mâche salad and dressed with a lemon dressing.

A word from the chef:
When cleaning the squid don't cut it open, to preserve their body sac.

Stuffed onions

Oignons farcis

Ingredients for 4 servings
2 cups rice pilaf − 2 large Spanish or sweet onions
2 oz. (60g) truffles, *brunoise*
2 cups brown juice from a roast − grated truffle

Prepare the rice pilaf.

Cook the onions, without peeling them, about 1 hour in a 300° oven.

Peel the onions and cut them in half horizontally.

Carefully remove the concentric layers, which will serve as many small bowls of varying sizes, to be filled with the stuffing you are about to make. Reserve the center of the onions.

Set aside the number of "bowls" you want to fill and chop the remainder, with the center.

Mix equal weights of chopped onions and pilaf and add the finely diced truffles.

Fill the onion bowls with this stuffing.

Braise gently in the 300° oven with brown juice from a roast.

If you'd like, sprinkle with grated truffle just before serving.

A word from the chef:

You can make an attractive presentation by arranging the stuffed onions by size, from largest to smallest.

Stuffed peppers

Poivrons farcis

Ingredients for 4 servings
3 cups rice pilaf − 8 small red peppers per serving
½ lb. (225g) slab bacon − 3 oz. (80g) truffles, finely chopped
½ cup truffle-flavored vinegar

Prepare the rice pilaf.

Remove the tops of the peppers and take out the seeds and the white ribs.

Cook the peppers, on a sheet tray in a 450° oven.

Blanch the slab bacon for 3 minutes in boiling water to remove the salt.

Cut it into tiny strips and brown it until crispy in olive oil.

Remove the peppers from the oven, cover them with a cool cloth, then peel them carefully.

Mix the pilaf with chopped truffles and the bacon, which will pick up some moisture from the rice but remain crunchy.

Fill the peppers with the rice stuffing. Heat in a 350° oven.

Serve with a truffle-flavored vinaigrette.

A word from the chef:

The blanched bacon must be thoroughly browned. In addition to its taste, it will add a bit of crunch to the stuffing.

Stuffed potatoes with truffles

Pommes de terre farcies

Ingredients for 6 servings
6 baking potatoes – 6 tbsp. butter – salt – freshly ground pepper
2 oz. (60g) truffles, sliced – 6 tbsp. clarified butter

Wash and brush the potatoes.

Cook them in their skins, in a 400° oven, turning them from time to time.

Cut the potatoes in two lengthwise. Carefully scoop the flesh out of the skins. Mash the flesh with a fork and mix in some good butter. Add a hefty dose of salt and pepper, then the sliced truffles.

The resulting purée should be somewhat soft, thanks to the butter.

Put the purée in a pastry bag and fill the skins in a nice pattern.

Brush with clarified butter.

Brown very briefly under the broiler.

A word from the chef:

When cooking potatoes in the oven, spread a layer of large salt crystals over a tray with sides.
You can serve the potatoes with grilled meat or present them with other stuffed vegetables.

Poultry

Guinea hen roasted in a bladder

Pintade en vessie

Ingredients

1 pork bladder from a specialty butcher — 1 large truffle or 2 small ones
1 guinea hen, free range — 4 slices sandwich bread
3 cups (¾ liter) milk — salt — freshly ground pepper — 3 small carrots
3 stalks of celery (white part only) — 2 leeks — 1 bay leaf
a few sprigs of thyme — white chicken stock — butter

Although the butcher should provide you with a bladder that's ready for use, rinse it to be absolutely sure.

Then slice a part of the truffle into thin strips. Cut the trimmings and the remaining truffle.

Using a knife with a sharp point, make several incisions in the skin of the guinea hen, lift the skin away from the flesh a bit, and insert the strips of truffle.

Make incisions on both the thighs and breasts.

Set the guinea hen to one side to allow the flavor of the truffles to permeate the flesh.

Make a light stuffing by dipping the bread in the milk, squeezing out the excess moisture and mixing with the rest of the truffle, which has been cut into small dice. (Reserve a small amount of truffle.)

Salt and pepper to taste. Stuff the guinea hen.

Then put the whole bird into the bladder and tie the ends of the bladder tightly so no fluids can escape.

Prepare the vegetables, washing them and tying the leeks in a bunch.

Place the guinea hen in a casserole with the vegetables and a bouquet garni of bay leaf and thyme.

Add the white chicken stock, all but covering the bird. Cover and cook very slowly at a simmer 200° for 1 hour. After 45 minutes remove the vegetables, drain them thoroughly, then cut them into small cubes, and quickly sauté them in a nonstick frying pan over high heat with a little butter.

Remove the guinea hen from the court bouillon and keep the precious broth for future use.

Although it's not general practice, cut open the bladder with scissors and carve the guinea hen at the table so your guests can enjoy the aroma that is released.

Serve with the vegetables, which should be mixed at the last moment with the remaining bits of truffle.

A word from the chef:

It takes some skill to keep a guinea hen from drying out in the oven. This particular cooking method retains all the delicacy of the hen after it is "dynamited" by the aroma of the truffles.

Truffled chicken breasts

Suprêmes de volaille truffés

Ingredients for 4 servings
1½ oz. (40g) truffles – 4 boneless chicken breasts, skin on
2 oz. (60g) butter – I cup crème fraîche – salt
freshly ground pepper – 12 oz. (350g) fresh pasta

Cut one truffle into bâtonnets.

Prick holes in the breasts with a pointed knife and fill them with the truffle sticks.

Julienne the other truffle and reserve.

Brown the breasts in butter in a heavy hot skillet.

Add crème fraîche to the level of the breasts. Season with salt and pepper.

Reduce the heat and cook very slowly.

In the meantime, prepare the pasta to serve with the sauce from the chicken breasts.

Finish the dish with a sprinkle of julienned truffle.

A word from the chef:
This dish is served with fresh pasta in a shape you like the most.

Chicken cream with truffles

Volaille à la crème de moutarde truffée

Ingredients for 4 servings

One 3½ to 4lb. (1,6–1,8 kg) chicken, free range – 2 tbsp. truffle oil
2 shallots, chopped – 7 oz. (200g) white mushrooms
½ cup dry white wine – 2 cups (½ liter) heavy cream – salt
freshly ground pepper – 1 cup brown sauce with 1 tbsp. mustard
fresh truffled pasta

Cut the chicken into 4 parts (breast with wings and legs).

Keep the backs to make a stock that will be of use on another occasion.

Using the truffled oil, cook the shallots until transparent, add the cut chicken, then the mushrooms.

Next deglaze with white wine and reduce, by three fourths lowering the heat. Stir in the heavy cream and add water enough to bring the level of the liquid even with the top of the chicken. Season with salt and pepper.

Cook approximately 15 minutes, depending on the size of the chicken pieces.

Remove the breasts and keep them warm. Let the legs cook for another 6–8 minutes.

Remove the legs and the mushrooms.

Reduce the sauce until it turns creamy. Add the brown sauce and blend it in thoroughly. Whisk to keep from boiling.

Pour the sauce over the meat and serve hot!

Accompany this dish with fresh pasta made with truffles.

A word from the chef:

In this recipe, truffles are not physically present; you won't be biting into them.
However, their taste in the oil is there to bolster the delicate flavor of the chicken and thereby create a subtle balance between the delicate flavors of the meat, cream, and mushrooms—and between the warmth of the truffles and the spice of the mustard.
Select a young free-range chicken or a capon for this recipe.

Meats

Filet of beef tenderloin

Chateaubriand truffé

Ingredients for each serving
1 beef tenderloin, 8 oz. (225g), 1½ in. (4 cm) thick − butter
¾ oz. (25g) truffles, minced − large French fries cooked in lard
1 salad with truffle vinaigrette

Preheat the oven to 400–425°. While the oven is heating up, sear the tenderloin in a heavy frying pan over high heat in foaming butter. As soon as the crust forms on one side, turn the tenderloin over. Keep the meat very rare.

Place it on a cutting board and cut in two, slicing horizontally to create a pocket. Put the minced truffles between the 2 slices of meat.

Carefully close the sandwiches and put them in the hot oven to finish cooking. The meat should go from extremely rare to medium rare.

Serve with large, crunchy French fries, and don't forget the salad.

Steak, fries, and salad—the typically French bistro meal, but improved a bit to be sure!

A word from the chef:
For joyful carnivores.

Veal chops with a truffled pouch

Côte de veau en poche à truffes

Ingredients for 4 servings

caul − 4 thick veal chops, 10–12 oz. (300–350g), 1½ –2 in. (4–5 cm)
salt − freshly ground pepper − 3 oz. (80g) truffles, finely chopped
¼ lb. (115g) butter − ½ cup white veal or chicken stock

Soak the caul in cold water.

Lay the veal chops flat on a cutting board and create a pouch by cutting them horizontally with a sharp, thin-bladed knife and make a pocket.

Sprinkle salt and pepper inside the pockets and fill them with the finely chopped truffles.

Wrap the veal chops with a piece of caul, just enough to wrap and close them. Cook the chops in butter.

Deglaze the pan with white stock. Allow the juice to reduce a bit, then pour it over the chops.

Your guests will have a nice surprise when they cut the meat, for the aroma of the truffles will quickly fill the room.

Veal chops studded with truffles

Côtes de veau cloutées

Ingredients for 4 servings
1½ oz. (40g) truffles − 4 thick veal chops, 10–12 oz. (300–350g),
1½–2 in. (4–5 cm) − butter − salt − freshly ground pepper
1 cup veal or chicken broth − 1 lemon, juice only

Cut the truffles into bâtonnets.

Make small incisions in the veal chops and fill them with equal amounts of truffle.

Cook the chops in butter until they are nicely browned.

Season with salt and pepper.

Reserve the chops, keeping them warm under a sheet of aluminum foil after sprinkling them with a few drops of lemon juice.

Deglaze the pan with the broth and reduce by half.

Remove from heat, whisk in butter in small pieces and decorate with any extra bits of truffles.

Pour the sauce over the chops.

You might accompany this with brussels sprouts and salsify.

A word from the chef:
Select the best of milk-fed veal.

Lamb loin wrapped in vegetables and caul

Noisettes d'agneau en crépinettes farcies

Ingredients for 6 servings
1 medium carrot − 1 stalk of celery − 3 cloves garlic − 1 medium onion
4 shallots − salt − freshly ground pepper − 1 piece large caul
3 lamb loins, fat trimmed − about 4 oz. (115g) truffles
4 tbsp. (60g) butter

Chop the carrot, celery, garlic, onion, and shallots very fine. Put them in an oven dish with a piece of damp parchment paper on top and cook them in the oven, covered, for 2 hours at 200 or 250°. They will turn completely to mush at this point. Season with salt and pepper.

Meanwhile, soak the caul.

Prepare and remove the fat from the lamb loins.

Julienne the truffles. When the vegetables are soft, flavor with the julienned truffles and then spread the lamb loins on both sides with this fine paste of vegetables and truffles.

Wrap the cutlets with the caul and cook the lamb loins in a frying pan with a little butter.

This dish can be quite extraordinary with Macaire potatoes (page 95).

Calf's liver piccata on a truffled onion compote

Piccata de foie de veau
sur compote d'oignons truffée

Ingredients for each serving
3½ oz. (100g) calf's liver − 1 or 2 red onions (depending on size)
1 tbsp. butter − 1 sprig fresh thyme − salt − 2–3 oz. (60–80g) truffles
½ cup dry white wine − flour, for dredging − freshly ground pepper
1 tbsp. balsamic vinegar

Ask your butcher to cut the liver into very fine, small scallops.

Cut the onions into long, very thin slices and cook slowly with a chunk of butter and the thyme. Salt the onions and wait patiently while they cook down to mush.

Meanwhile, cut the truffles into very fine dice.

Add a bit of white wine to the onion compote.

Dredge the liver scaloppine in seasoned flour. Then cook them quickly in a sauté pan with butter.

Take care to keep them a little rare and pink; you don't want them to become tough.

Remove the liver and keep warm. Deglaze the pan with balsamic vinegar.

Sprinkle the onions with truffles and heat briefly.

Serve the liver on the truffled compote and drizzle with the deglazed juice.

A word from the chef:

Use onions that have a fairly strong taste; they'll lose some sharpness during cooking and they'll give less sugar. Be sure to salt the onions before you cook them to release the water they contain.

Oxtail ragoût

Ragoût de queue de bœuf

Ingredients for 6 servings
1 oxtail approximately 6 lbs. (2½ kg) − 3 carrots
3 small turnips − 3 small stalks of celery
1 leek − 1 bouquet garni − 1 onion, studded with cloves − salt
endives − 7 oz. (200g) truffles, chopped

Begin by soaking the oxtail in water for 12 hours.

When that's done, cook it like a stew with the carrots, turnips, celery, and leek, the bouquet garni, the onion, and water to cover.

Add a little salt, but not too much, because you're going to reduce the broth later.

Blanch the endive leaves at a rolling boil for 3 minutes.

Remove the the oxtail from the broth. Drain the vegetables and reserve them. Strain the broth and reduce it until it is almost a glaze. Meanwhile, carefully remove the bones from the oxtail. Chop the vegetables and add them to the picked meat shreds.

Don't forget to remove the cloves. Then add the truffles and the reduced broth. Thanks to the vegetables and the broth, you'll have a very soft, smooth stuffing. Drain the endive leaves and use them to line six 3-in. (7,5 cm) ovenproof ramekins. Then fill the lined molds with the stuffing. Cook briefly in a 350° oven to reheat the meat.

A word from the chef:

You can also put this stuffing in phyllo dough and then cook in a hot 425° oven. That way you'll have a contrast between the crunch of the phyllo dough and the smoothness of the stuffing. Serve with a vinaigrette made from olive oil. I find that a vinaigrette goes nicely with the taste of the truffles.

Lamb tripe

Tripes d'agneau

Ingredients for 6 servings
4½ lbs. (2kg) tripe – 2 onions – 2 carrots – 3 stalks of celery
thyme – bay – 1 bottle dry white wine – peppercorns
3½ lbs. (1.5kg) baking potatoes – 9 oz. (250g) truffles, chopped

This is a simple recipe, but it takes a long time to cook.

If you know a good specialty butcher, he will sell you tripe that is perfectly cleaned and blanched.

If not, wash the tripe thoroughly and blanch it.

Prepare the vegetables by cutting into a medium dice and put the onions, carrots, celery, thyme, and bay in a large, heavy pot with the bottle of white wine and an equal amount of water.

Add a few peppercorns (but not too many) and cook for 6 full hours.

Prepare the potatoes to accompany the tripe. Steam them.

Chop the truffles somewhat roughly.

When the tripe is cooked, toss the truffles into the pot. Stir gently but thoroughly and put the cover back on.

Remove the pot from heat and allow to infuse for 3 minutes, not more, and serve very hot with the potatoes.

Lovers' tartare

Tartare des amoureux

Ingredients for each serving
5 oz. (150g) prime beef or filet mignon – 1½ oz. (40g) truffles
1 egg yolk – olive oil – salt – freshly ground pepper
endive salad with truffle-flavored vinaigrette

Slowly mince the meat.

Chop the truffles, but not too finely!

Mix together. Make an indentation for the egg yolk. Season with salt and pepper, add a little olive oil, and mix on the plate.

Present with a crisp endive salad dressed with truffle-flavored vinegar.

Then put on some soft music and . . .

A word from the chef:

Avoid using a food processor, because it alters the flavor of the meat. A good meat grinder will treat the prime beef with respect.

Sauces

Celery root sauce

Sauce au céleri

Ingredients for about 2 cups (½ liter)
1 lb. (450g) celery root − 1 lemon − 1 qt. (1 liter) white stock − salt
pepper − 4 oz. (115g) truffles, chopped

Begin, of course, by peeling the celery root. Then cut it into large cubes. Sprinkle it with lemon juice to keep it white.

Then cook it in the white stock.

Pour the celery root and stock into a blender or food processor to blend them together and make a sauce. If the sauce is too thick, thin it with white stock. If it is too thin, reduce it to arrive at the desired consistency.

Season with salt and pepper. The pairing of celery root and truffles needs no other seasoning, because the celery root is very strong.

Add the chopped truffles to the sauce while it is hot.

A word from the chef:
The combination of truffles and celery root is very agreeable.
You can can cover chicken breasts with this sauce and cook them in alumiunum packets.
Veal cutlets or chops also go nicely with this mixture.
Finally, you can surround a piece of grilled meat with leeks and carrots in an attractive way and cover them with this sauce.

Truffled mustard sauce

Sauce à la moutarde truffée

Ingredents for about 2 cups (½ liter)
2 cups (½ liter) heavy cream − ⅓ oz. (10g) cornstarch
salt − pepper − 1 tbsp. truffled mustard

Boil the cream over very high heat for a minute or two. Thicken with the cornstarch. Add salt and pepper.

Stir a few times to allow the ingredients to blend.

Incorporate the mustard, taking care to prevent the sauce from boiling.

You can, if need be, thin the sauce with a little water. This sauce can do wonders for an andouillette sausage.

A word from the chef:
To accompany grilled or poached fish.
To turn sea bream royale *into sea bream* imperiale.

Port sauce with truffles

Sauce porto aux truffes

Ingredients for about 2 cups ($\frac{1}{2}$ liter)
2 shallots, very finely chopped – butter – 1 bottle of red Port
2 cups ($\frac{1}{2}$ liter) rich veal stock or thickened broth – salt – pepper
$2\frac{1}{2}$ oz. (70g) truffles, roughly chopped

First, gently sauté the shallots with a chunk of butter. Before they turn white, add the Port and reduce it until it's nearly dry. You don't want it to carmelize, but it must be greatly reduced so the juices are concentrated.

At this point, add the veal stock and allow it to cook at a low boil for a full 15 minutes.

A little fat will float to the top; skim it off. To make good sauces, you must remove what rises to the surface during cooking.

Stir in just over $1\frac{1}{2}$ oz. (50g) of butter cut into small pieces.

Add salt and pepper.

Put the chopped truffles in another container and pour the sauce over them.

The truffles will be cooked by the heat of the broth and nothing else.

A Périgueux sauce

Une des sauces Périgueux

Ingredients for about 2 cups
4 oz. (110g) cooked ham − 1 onion, minced − 1 shallot, minced
1 bottle of red wine − 2 cups (½ liter) Port − 1 bouquet garni
1 tbsp. lard or oil − 1 tbsp. flour − 1 qt. (1 liter) brown stock − salt
freshly ground pepper − 3 oz. (90g) truffle

Cut the ham into small pieces and brown with the minced onion and shallot.

Add the wine and the Port and reduce by half with the bouquet garni.

Meanwhile, make a roux with the lard or oil and the flour.

Cook until it turns a bit brown.

Add the roux to the wine and the brown stock. Cover and cook slowly for 1 hour. Remove the bouquet garni.

Season, but go easy on the salt because the ham is salted.

Chop the truffles, mix them into the sauce, and cook the sauce for 2 more minutes.

A word from the chef:

There are many ways of making a Périgueux sauce.
Here's one that you can prepare easily at home.
The lard gives this sauce something of a rustic touch.

Vegetables

Artichoke flan

Flan d'artichauts

Ingredients for 6 to 8 servings
7 oz. (200g) artichoke bottoms − 1 lemon − 1 bay leaf
1 sprig of rosemary − 7 eggs
2 cups (½ liter) heavy cream − 2 oz. (60g) truffles

Prepare the artichokes. Sprinkle them with lemon juice if they must wait to be cooked.

Cook them with a bay leaf and the rosemary.

Then place them in a food processor or blender with 3 whole eggs and 4 yolks. Add the heavy cream and mix again.

Strain through a fine conical strainer (china cap).

Slice the truffles into a fine julienne.

Add them to the flan. Preheat the oven to 300°F.

Pour the mixture into either a 9-inch flan mold or small individual ramekins. Cook slowly and gently in the oven in a water bath.

It's the slow coagulation of the eggs that will make your flan a perfect success.

A word from the chef:

Use the rosemary with caution. Its taste is very powerful and will spread all through the flan. It can mask the other flavors.
A little is enough of this herb.

Potato galette with truffles

Galette de pommes de terre aux truffes

Ingredients for 4 servings
1 medium truffle about 2 oz. (60g) − 3 russet potatoes
3 oz. (80g) butter − salt − pepper

Brush and peel the truffles (keep the peelings) and cut them into fine strips.

Peel the potatoes then grate them over a plate.

Clarify the butter over low heat and skim thoroughly.

Brush the butter over the inside of ovenproof molds.

Put layers of grated potatoes in the molds, separating the layers with a strip of truffle. Salt and pepper.

Baste with clarified butter and place in a very hot oven (450°F) for 15 minutes.

Present these ramekins as an accompaniment to lamb chops that have been browned on the grill and decorated with a few sprigs of fresh marjoram.

Cardoon gratin

Gratin de cardes

Ingredients for 2 servings
I large cardoon − lemon juice − I cup flavorful chicken or roast juice or
crème fraîche − as many truffles as you please
salt − pepper

Wash, peel, and cut the cardoon into matchsticks.

Then cook in water to which you've added some lemon juice.

Drain the cardoon and put it under cold running water to cool it. Press it carefully and drain it once more.

Next, preheat the oven to 400°F. Put the cardoon in a gratin dish and pour over the best chicken or other meat juice that you have kept for this sort of use. Add as many truffles as you would like.

If you don't have any roast juice, you can use crème fraîche spiced with a few drops of glace de viande. Place in the oven.

If you're fond of cheese, select the best parmesan you can find (not pregrated) and grate it over the cardoon just before you put it in the oven.

A word from the chef:

Cardoons should be cooked in plenty of water. You'll need a large pot for the purpose.

Gratin of asparagus tips with truffles

Gratin de pointes d'asperges aux truffes

Ingredients for 6 servings
5–8 asparagus tips (depending on size) per serving
3 oz. (80g) melted butter – ½ cup (70g) flour – 1 qt. (1 liter) chicken
stock – salt – pepper – nutmeg – ½ lemon
7 oz. (200g) truffles, julienned – 2 eggs – butter

Cook the asparagus tips in salted water. They should not be crowded.

Make a velouté with the chicken stock.

To do that, first make a white roux with melted butter and the flour. Cook in a saucepan, removing from heat before the roux begins to take on color.

Add the chicken stock and cook slowly and gently. Season with salt, pepper, a little nutmeg, and a few drops of lemon juice.

Thoroughly drain the asparagus tips. Then put them in a gratin dish and cover them with truffles.

Preheat the oven to just over 200°F.

Remove the velouté from the heat and incorporate 2 egg yolks, beating gently with a whisk.

Cover the asparagus and truffles with the velouté, dot with butter, and place in the oven for 1 hour. The velouté will cook a little like a flan.

Ten minutes before serving, increase the heat and place under the broiler to brown the surface.

A word from the chef:
The first asparagus arrives in the market before the last truffles disappear.

Cabbage packets with truffles

Paupiette de chou aux truffes

Ingredients for 4 servings
4 large cabbage leaves – 3½ oz. (100g) ham – 3½ oz. (100g) veal breast
1 small glass of Armagnac – 3 oz. (90g) mushrooms – 1 shallot – salt
pepper – 2 tsp. crème fraîche – 4 truffles, each weighing 1–1½ oz.
(30–40g) – 1 cup of strong stock – 1 small onion – ½ carrot
1 stalk of celery – 1 tbsp. butter

Blanch the cabbage leaves, wring them out or give them a spin in a salad spinner, and reserve them.

Chop the ham and the veal and pour the Armagnac over them.

Chop the mushrooms and the shallot very fine and gently brown them along with the ham and veal.

Add salt and pepper and 1 teaspoon of crème fraîche. Stir without allowing the mixture to come to a boil.

Wash and brush the truffles.

Spread the cabbage leaves on the table and put some of the stuffing (the meat and mushrooms) in the center of each leaf. Push a truffle into each mound of stuffing, making sure that the truffle is completely covered.

Preheat the oven to 350°F.

Fold the cabbage leaves over the stuffing and set them in a gratin dish folded side down. Cover halfway with the stock, to which you've added the finely diced onion, carrot, and celery. Cook for 30 minutes, basting frequently.

Remove from the oven and keep hot. Thicken the remaining broth with butter.

Enliven the sauce with extra bits of truffle and serve.

Macaire potatoes with truffles

Pommes Macaire aux truffes

Ingredients for each serving
2 large potatoes – as many truffles as you like
2 oz. (60g) clarified butter – salt – pepper – chervil leaves

Bake the potatoes in the oven.

Meanwhile, clarify the butter. Prepare the truffles and cut them first into thin slices and then into a fine julienne.

Make a purée with the flesh of the potatoes and 2 oz. (60g) of butter for 2 potatoes (no milk or cream).

Add salt and pepper.

Gently but thoroughly stir the julienned truffles into the purée.

Form small pancakes about ½ inch thick. Brown both sides of these pancakes in clarified butter over high heat.

Decorate with chervil leaves and serve as an accompaniment to a grilled meat.

Cauliflower purée with truffles

Purée de chou-fleur aux truffes

Ingredients for 4 servings
¼ lb. (115g) potatoes – 1¾ lb. (800g) cauliflower – 1 qt. (1 liter) milk
salt – pepper – 2 oz. (60g) butter – ¼ cup crème fraîche
3 oz. (90g) truffles

Peel the potatoes and cut them into quarters. Prepare the cauliflower and cook with the potatoes in milk over low heat.

When the vegetables are cooked, put them through the food mill. Season with salt and pepper to taste.

Dice the butter and turn it into the purée with a rubber spatula, one or two pieces at a time. Then thin with the crème fraîche to the desired consistency.

Keeping the purée warm, grate the truffle over the top. (Use the disk on the food mill that's used to grate cheese to get a fine texture.)

Turn the grated truffle into the purée, making sure the truffle is thoroughly incorporated.

Serve as an accompaniment to veal cutlets, scallops of foie gras, or roasted pigeon.

A word from the chef:

Use a good old hand grater. To serve as an hors-d'oeuvre, make quenelles (dumplings) from the purée, shaping them with a soup spoon. Garnish with a thin slice of truffle and sprigs of chervil.

Shredded russet potatoes

Râpée de bintjes

Ingredients for each serving
1 russet potato (2 if small) − 2 oz. (60g) butter
1 oz. (30g) truffles

Wash and peel the potatoes.

Clarify the butter over low heat.

Chop the truffles and reserve them. Grate the potatoes and separate the pulp into two piles.

Put the clarified butter in a nonstick frying pan. Add a layer of potatoes followed by the truffles and then a second layer of potatoes. Brown the resulting pancake on both sides, turning it with the aid of a large flexible spaluta or a large flat plate.

A word from the chef:

To flip the pancakes, you'll need a large flexible spatular or a flat plate of the same diameter as the frying pan.

Terrine Stendhal

Stendhal Terrine

Ingredients for 8 servings
3 red bell peppers – salt – pepper – 1 truffle weighing 3½ oz. (100g)
mâche salad – truffle-flavored vinaigrette

Here's a nice dish that offers interesting contrasts in color and texture.

Put the red peppers in a very hot oven 450° to 500°F to blacken the skins quickly.

Peel the peppers and wipe them meticulously; you don't want any of the burned skin to adhere to the flesh.

Season with salt and pepper.

Cut the truffle into thin slices.

Select a terrine that is neither very large nor very tall, to make several layers.

Place a layer of peppers in the bottom, then a layer of truffles. Alternate layers in this fashion, ending with a layer of peppers. Set a second terrine dish on top of the peppers and weight it down with one or two large heavy cans (or another weight).

Place in the refrigerator for 24 hours.

To serve, slice and present on a plate, surrounding the red and the black with the green of the mâche salad, which should be dressed with truffle-flavored vinaigrette.

A word from the chef:

To make your peppers easy to peel, remove them from the oven with their skin bubbling and cover them immediately with a cold, dry cloth. The steam will loosen the skin.
This terrine is named after Stendhal only because of the colors. There is no chartreuse in this recipe—not even de Parme!

Rice and Pasta

Homemade pasta with truffles

Pâtes maison aux truffes

Ingredients for 4 to 6 servings
2 cups (300g) flour – 3 eggs – ¼ cup crème fraîche
2 small truffles about 2 oz. (60g) – 1 shallot – salt – pepper
vegetable oil

Make a circle with the flour. Break 2 eggs into the center. Add 1 teaspoon of crème fraîche. Mix together and knead for 10 full minutes.

Form the dough into a ball and put it to one side under a damp cloth for 1 hour.

Pull the ball into 3 or 4 pieces. Dust your work surface and rolling pin with flour and roll the dough out as thin as possible.

Cut into thin strips with a knife or a ravioli wheel.

Give the pasta dough a fun shape.

Allow to sit between sheets of clean cloth until the next day.

Preparation of the sauce:

Cut the truffles into very small dice.

Mince the shallot fine and sauté it in butter. Add the diced truffles. Stir gently for 3 minutes.

Add the remaining crème fraîche and blend thoroughly. Season with salt and pepper.

Remove from heat and thicken with the yolk of the remaining egg. Keep hot without cooking further.

While preparing this sauce, put a large pot of salted water on to boil. Add a few drops of oil to the boiling water. Then drop the pasta into the water and cook at a rolling boil until al dente.

Drain the pasta as soon as it's cooked, mix with the sauce in a salad bowl, and serve pronto.

A word from the chef:
Prepare the pasta one day, cook it the next.

Rice pilaf

Riz pilaf

Ingredients
2–3 oz. (60–80g) whole, long-grain rice per serving – 1 onion, chopped
butter – olive oil – white stock

Sauté the rice with the onion in an ovenproof pot with equal amounts of butter and olive oil.

Add twice the amount of white stock.

Cook in the oven for the amount of time specified on the rice package.

Fluff the rice thoroughly with a fork while adding some butter a little at a time.

This rice can be used as a garnish to set off a meat or fish dish in a discreet way, or as a stuffing in certains dishes.

A word from the chef:
Cooking dictionaries often spell pilaf pilaw, as it was known in the time of Gouffé or Escoffier, great French chefs of the 19th and early 20th centuries.
An ideal accompaniment for any dish.

Cheese

Stuffed camembert

Camembert fourré

Ingredients
1 camembert − strips of truffle

This is very easy to prepare. Everything depends on the quality of the cheese.

Select a ripe camembert, but pay close attention to how mature it is. The camembert must be firm in the center, not runny. You can buy 2 camemberts and use one to check maturity.

When the cheese is ready, scrape the top and remove the hard crust around the edge and on the top and bottom.

Then cut the camembert in two, slicing horizontally through the center. Put several slices of truffles inside.

Wrap the cheese in plastic and set it in the bottom of the refrigerator for 1 or 2 days.

Serve on toasted country-style bread and keep warm in a napkin.

Open the best red Châteauneuf du Pape you can find and enjoy a great culinary event made possible by combining the simplest of ingredients.

A word from the chef:
Camembert country is quite distant from truffle country. Nevertheless, these two distinct flavors complement each other nicely.

Toast with goat cheese from the Vaucluse region

Tartine de chèvre du pays de Vaucluse

Ingredients
slices of country-style bread – very fresh goat cheese
fine sea salt – pepper – a few slices of truffles
a cruet of olive oil

Toast the bread, cut it into canapés (squares without crust), and keep warm.

Cut the goat cheese into thin slices. Alternate slices of cheese and truffle (to make a good contrast in color) on the canapés.

Season with salt, and don't hesitate to add pepper.

Pour a few drops of olive oil on each canapé. Use a very fine olive oil, one from Nyons, Aix, or Les Baux (all olive oils from Provence).

These canapés make a perfect hors-d'oeuvre.

A word from the chef:

A few slices of goat cheese, a bit of fine sea salt, a few drops of olive oil, and a thin slice of truffle can give you a foretaste of happiness.

Desserts

Sweet beignets with truffles

Beignets de truffes

Ingredients

as many truffles as you like
Beignet dough:
1½ cups (225g) flour − 1 tsp. oil − 1 cup beer
1 cup water − pinch salt − 2 tsp. sugar − 2 egg whites
Caramel:
¾ cup (190g) sugar − ¼ cup water
vegetable oil, for frying

In a bowl, prepare the sweet beignet dough with the flour, oil, and beer.

Allow it to sit at least 30 mins. before folding in the whites of 2 eggs, which you've beaten into stiff peaks.

Make a very simple caramel with the sugar and water.

Cut the truffles into somewhat thick strips.

Heat oil in a frying pan.

Fry the beignets over high heat and drain them on paper towels.

Then dip them in caramel and eat them!

Ice cream

Crème glacée

Ingredients
2 cups (½ liter) milk − ½ vanilla bean
8 egg yolks − 1¼ cups (300g) sugar − 1 cup crème fraîche
7 oz. (200g) truffles, minced

Put an empty bowl in the refrigerator. Then put a pot on the stove and boil the milk with the vanilla bean.

Make a vanilla sauce (*crème anglaise*) by beating the egg yolks with a generous cup of the sugar. Add the milk and cook over low heat. The cream is ready when it coats a spoon.

Remove the vanilla and allow the cream to cool.

Meanwhile, take the cold bowl from the refrigerator and put the crème fraîche and the remaining sugar in it. Beat them together to make a crème chantilly.

When the cream has cooled, gently fold in the chantilly and the minced truffles.

Put the finished cream into one or more molds and place in the freezer.

For a memorable experience, serve with a Beaume-de-Venise that has almost maderized or an old Loupiac or Sauternes that you've forgotten in your wine cellar.

Apples Arphrin

Pommes d'Arphrin

Ingredients for each serving
1 apple – powdered sugar – a few drops of lemon juice
¾ oz. (20g) truffles, cut into very small dice – 2 tbsp. cognac
2 eggs – butter

Peel the apple and grate it.

Coat with powdered sugar by adding the sugar a little at a time while stirring. Add a few drops of lemon juice to keep the apple from turning brown. Then add the finely chopped truffles and the cognac.

Blend everything together thoroughly.

Beat the whole eggs and mix with the apples.

Cook the mixture as you would an omelet in a little bit of butter. Turn the omelet with a flexible spatula to cook the other side.

A word from the chef:
Turn this omelet with a flexible spaluta or a large plate.
Use a nonstick pan.

Ratafia of love

Ratafia d'amour

Ingredients
5¼ oz. (150g) truffles — 2 vanilla beans
1 qt. (1 liter) Poire William distilled fruit alcohol, very strong
and clear (firewater!) — 7 tbsp. (100g) sugar — 1¼ cups water

Clean the truffles. Chop them roughly but delicately.

Break the vanilla beans in two along their length. Soak the truffles and the vanilla in the 2 tbsp. Poire William distilled fruit alcohol for 20 days.

At the end of that time, remove the vanilla and strain the Poire William distilled fruit alcohol, squeezing the liquid out of the truffles.

Reserve the truffles for future use (with wild game, for example).

Dissolve the sugar in the water and add the Poire William distilled fruit alcohol. Pour into bottles and seal them hermetically.

Allow this to sit for 1 month in the wine cellar before consuming.

A word from the chef:
There's a well-known story of a lady who enslaved two ministers of government with this ratafia. Her descendants are still very rich.

Fruit salad

Salade de fruits

Ingredients per serving
½ grapefruit – 1 orange – ½ passion fruit – brown sugar, to sprinkle
¾ oz. (20g) truffles per serving – 1 tbsp. Grand Marnier

Peel the citrus and the passion fruit and arrange them attractively on a plate. You can make a very pretty composition with passion fruit, oranges cut into slices or quarters, and grapefruits.

Sprinkle the fruit very lightly with brown sugar.

Grate truffles over each plate, speckling the flesh of the fruit with black.

Drizzle with Grand Marnier. The brown sugar and the liqueur will temper the acidity of the fruit.

A word from the chef:

All citrus go well with truffles.
Grate the truffles just before serving.

Hot soufflé

Soufflé chaud

Ingredients for 6 servings
6 egg yolks – 1 generous cup (250g) sugar – 2 cups (½ liter) milk
½ cup (75g) flour – ½ vanilla bean – 4 egg whites
7 oz. (200g) truffles, chopped

First make a pastry cream by beating the egg yolks with the sugar.
Boil the milk and add it to the egg yolks. Then gently fold in the flour.

Beat the egg whites to stiff white peaks. Thoroughly blend them into
the custard, along with the chopped truffles. Preheat the oven to 350°F.

Spoon the mixture into one or more soufflé molds and place in the
oven. After cooking, raise the temperature in the oven to 400–450°F to
finish the soufflé.

Frozen whisky soufflé

Soufflé glacé au whisky

Ingredients
7 oz. (200g) truffles — 1 small glass of whisky
1 generous cup (250g) sugar — water — 6 egg whites — 6 egg yolks

Cut the truffles into very small dice and soak them in the whisky for 12 hours.

Then make a sugar syrup by boiling the sugar with a little water.

While the syrup is cooking, beat the egg whites into peaks of snow.

The syrup is ready when it reaches 250°F on a candy thermometer or, more simply, when the syrup forms a thread when tipped from a spoon.

At that exact moment, fold in the egg yolks and beat until the mixture has cooled completely.

Then blend in the beaten egg whites with a rubber spatula, also adding the truffles and the whisky.

Fill several molds with this mixture and place them in the freezer.

A word from the chef:

Use a single malt Scotch with a light peat taste, which will intensify the flavor and the aroma of the truffles.

Truffles with truffles

Truffes de truffes

Ingredients
8¾ oz. (250g) excellent bittersweet chocolate
½ cup crème fraîche – 3 oz. (90g) truffles, finely chopped
I cup the best cocoa powder that you can find

Make a ganache by gently melting the bittersweet chocolate and folding in the crème fraîche so that it is blended completely.

Blend in the truffles while the ganache is still warm; the heat will diffuse the flavor of the truffles all through the chocolate.

Next, make small balls or almond shapes with the chocolate and roll them in the cocoa powder.

Place in the refrigerator under a sheet of aluminum foil.

A word from the chef:
No one knows the origins of the word ganache.
All pastry chefs and cooks in France know the word and agree on its meaning, but no one knows who came up with it, not even M. Bocuse, to whom I put the question!

Frozen truffles

Truffes glacées

Ingredients for 8 truffle balls
16 egg yolks – 2 generous cups (500g) powdered sugar
1 qt. (1 liter) milk – ½ vanilla bean – 4 oz. (115g) truffles
cocoa powder

Make a vanilla sauce (*crème anglaise*) by whisking the egg yolks with the sugar.

At the same time, bring the milk with the vanilla bean to a boil.

Remove the vanilla and pour the milk over the egg yolk mix. Cook gently until the cream coats the spoon.

Allow to cool. Meanwhile, chop the truffles fine and fold them into the cream while it's still warm.

Put immediately in an ice cream maker.

Mold the cold, thickened cream into balls and roll them in the cocoa powder.

Place immediately in the freezer.

A word from the chef:
You'll need an ice cream maker.

Cassanova's wine

Vin de Casanova

Ingredients
1 bottle of white wine, such as Coteaux du Ventoux
2 tbsp. Poire William distilled fruit alcohol, very strong
and clear (firewater!) – 2 small truffles, diced
3 pieces of unrefined cane sugar – 6 sprigs of savory

Remove 1 glass of wine from the bottle.

Add the Poire William distilled fruit alcohol to the bottle, along with the truffles and the sugar. Put the savory in a cheesecloth pouch and tie it closed with a string.

Push the sachet into the bottle while holding on to one end of the string.

Recork the bottle, leaving the string dangling outside. Allow the bottle to sit for 28 days (one lunar cycle).

At the end of that time, uncork the bottle and remove the savory by pulling on the string.

Adjust the flavor by adding sugar if necessary.

Cork the bottle and seal with wax. Allow the wine to age if you can make yourself wait. You can add a tiny pinch of powdered cinnamon.

A word from the chef:

Make a cheesecloth pouch or sachet to hold the savory. Close the bag with string, leaving a 6-in. (15-cm) length of string trailing from the knot.
An alchemist would insist that the flavoring of the wine begin the first lunar quarter.
Served as an aperitif, this ratafia predisposes your guests to enjoy the meal to come.

The Gourmand's Truffle

The truffle, for the gourmand who is about to savor it, is first of all a sigh of comfort and contentment that can be translated as: "Finally, the great pleasure is here!" Then one or both of his nostrils twitch, and he lowers his eyes or closes them in order to hide the gleam emanating from deep in his pupils. With his eyes still lowered, he discreetly leans over his plate, inhales deeply while lifting his head, inhales again, and finally smiles at his dinner partners.

This involuntary and inevitable ritual is the chef's reward. He is gratified by his efforts and by the care he has taken to prepare this rare and exceptional product that nothing artificial has yet replaced.

The Truffle Hunter's Truffle

For the truffle hunter, the truffle is an object of desire, like a piece of gold for Balzac's Père Goriot. When, thanks to mysterious clues or the behavior of his dog, he knows there's a truffle at his feet, he stops, looks nervously around him, inspects the surrounding trees and bushes, listens for sounds, makes himself as inconspicuous as possible, crouches down, digs at the spot indicated by his dog, slips the truffle in his pocket, rewards the dog, fills in the hole, wipes away any evidence of digging, and moves on in silence, holding his pleasure in abatement.

It's only after returning home and ensuring that he is alone that he will admire the fruits of his plunder. He sniffs, scrapes, weighs, speculates, counts, and caresses each soil-covered ball as he removes it from his pocket, then hides his treasures before going out later to sell or trade them.

The explanation for this strange behavior is that a large truffle is worth the market price of a piece of gold.

Imagine that every day, thanks to a small dog with eyes that are full of devotion and love, you're able to go out and find 10, 15, 20, or more gold coins beneath the soil, and you'll understand the passion of the truffle gatherer, who sees more in this mushroom than its market value. For him, there is the joy of being one of the initiated, the joy of being the person who finds the precious gem that others have walked right by, the joy of being someone to whom the gods of nature, the woods, and the earth have revealed a great secret.

The Truffle for the Botanist

For the botanist, a truffle is a mushroom of which there are several species. These mushrooms belong to the genus *Tuber*, and the species include *Tuber melanosporum*, *Tuber aestivum*, *Tuber brumale*, *Tuber uncinatum*, *Tuber magnatum*, *Tuber moscatum*, etc. A truffle can be black; it may mature in summer and be white; it has a musky odor that can contain a hint of garlic, etc. Like all mushrooms, truffles create an underground network of filaments that are microscopic in thickness but that may be very long indeed. This network is called the mycelium, and to reproduce, the truffle produces spores.

What sets truffles apart from most other mushrooms is that they have an interdependent relationship with a tree, which plays a somewhat mysterious role in their development.

Botanists speak of a symbiosis between truffle and tree. The oak is the best known companion for truffles—the red oak (*Quercus robur*), but also the holly oak (*Quercus ilex*) and the pubescent oak (*Quercus pubescens*).

There are, however, several other kinds of trees that host the truffle mycelium among their roots: the filbert, the linden, the poplar, the elm, the beech, the sea or wood pines, Austrian pine or (*pin d'alep*), the ash—almost any tree that is able to produce, at low altitude, a large network of roots in a friable soil that is covered with humus.

The Truffle for the Researcher

For agricultural scientists, truffles are a big headache.

Think about it. No one has yet figured out how to produce truffles the way we grow tomatoes or raise chickens or rabbits.

You can buy shrubs whose roots have been inoculated with the truffle fungus. All you need to do is plant them in a suitable location (whatever that is) and wait at least eight or nine years after having watered, pruned, watched, watered again, and prayed that Mother Nature, by some miracle of which only she is capable, finally give you the joy of harvesting the first black jewels, and there is by no means any guarantee that your efforts will be rewarded.

You have to admit that for a research scientist who has many years of schooling behind him this uncertainty is enough to make him eat his hat out in frustration.

But that's the way it is with truffles, and maybe it's just as well. Truffles remain mysterious, something you can dream about a little bit.

The Truffle for the Cook

The truffle, for a cook, is a flavor, an aroma, a taste, a meal in itself, a powerful ingredient. It's one of the most precious and most noble things the earth can produce, and it's absolutely natural. But beware: A jewel such as this requires a little humility. Yes, truffles flavor the things they come in contact with, whether it's the clothing of the person who gathers them or eggs still in their shells, but that's no reason to put them in every sauce nor to sprinkle them over failed dishes. There are incompatibilities of nature and season that must be heeded. At the same time, there are natural alliances, harmonious relationships to explore, to discover, to understand, to intensify. Because they remain veiled in mystery, truffles lend themselves to experimentation, and like all precious gems, they fall victim of pretentiousness. It takes a lot of effort to study truffles and to treat them with the respect they deserve, but it's worth the trouble so that the gourmand still has things to discover, so that he can finally indulge in the rite of contentment over a dish where truffles are at work.

Since man has been cooking, truffles have been regarded as quasi-sacred and attributed with magical powers. Like everything that is considered sacred and mysterious, it has engendered some very curious rituals and odd behavior. There is, for example, the market where truffles are sold, or rather exchanged between connoisseurs in return for money. It is good form to appear not to want to buy them. It's also good form to appear not to want to sell them: "This year, there aren't any!" There haven't been any for a hundred years and more. The rarer something is, the more expensive it is and the more adroit, skillful, and cunning is the person who managed to find some despite all the misfortunes that have afflicted the region and the luckier the buyer who encounters a good gatherer.

And of course there is the price, the going market rate, which fluctuates like the price of shares in the trading pit at the stock market. Who sets it? No one knows! Word of mouth and rumor have a huge impact on setting the value of this Dow Jones Industrial stock.

But the price is the price. The buyer tries to negotiate it down; the seller strives to keep it high. Once agreement is reached, there's no more room for adjustment.

One day, a broker stopped by the restaurant. We had known each other for a long time. We came to agreement on quantity and price, and the deal was made. He weighed the truffles with care and calculated their value. The price came to 15 francs more than a round sum. I said to him: "Shall we round it off?" "Not on your life!" For more than half an hour, we argued. He did not want to knock 15 francs off the price. I gave in, and we drank champagne to celebrate the transaction. He paid 120 francs for the champagne with a smile, but he wouldn't give away 5 grams or even 1 gram of truffles. They're all like that. Generous with their money to show their friendship but greedy and secret when it comes to their cursed truffles.

Truffles have been prized by lovers of fine food for 4,000 years (since the Sumerians), and they were also banned at one point by the Church, because they were said to be an aphrodisiac. Their black flesh, their reputation as a sexual stimulant, the mystery surrounding their appearance in the soil gave some to believe they were the physical manifestation of the souls of the damned, which were blackened by their sins.

It used to be said in Provence that truffles are blacker than the soul of the damned.

Is the truffle an aphrodisiac? Does it sharpen one's sexual appetite? Opinions are divided. On this subject, here is a strange story that was told nearby Roumoules (Alpes de Haute Provence) until about 30 years ago, before the area was deforested, built up, and overrun with people.

The Fable of the Skeptic and the Truffle Digger

Once upon a time, there was a man who was a sorcerer, bonesetter, poacher, and truffle digger.

He didn't like, of course, for anyone to follow him into the oak woods or along the banks of the Colostre River, and he had an amazing knack for disappearing into his surroundings.

People went to see him to find a missing jewel or to get treatment for lower back pain, rheumatism, or a bad cold. He also treated animals. His dog was accused of being as much a sorcerer as his master, but he was just a good old basset hound and a precious and indispensable aid in finding truffles.

There was a man who had to leave the region as a boy when his parents died. He returned to live on a pension that allowed him to fix up the family home and lead a simple life of gardening and reading—over and over again—a collection of books he had brought with him in a large metal trunk.

Life in the city had converted him to the cult of reading and science and had made him a free thinker.

One day, suffering from painful lumbago, he went to see the bonesetter, who eased his discomfort. The sorcerer didn't want money for his services. So the man offered him a gift that seemed very precious: a book on botany with scientific sketches. The good bonesetter was all the more touched by this gift because he could not read.

Bit by bit a friendship, then a strong bond developed between the two loners.

The poacher never admitted that he didn't know how to read, but he confided that he never ate the truffles that he dug and sold or traded.

The free thinker had the good manners not to laugh when he learned that the truffle digger believed wholeheartedly that truffles were the physical desires of people who had died of concupiscence and that that explained why truffles were black.

Of an evening, after a brochette of thrush or a roasted partridge, they had lively discussions. One spoke of the harmless mushroom, the other produced proofs that supported his conviction: His dog had been seized with lust after nibbling on a small truffle, and the bitch on the neighboring farm gave birth to no less than 12 puppies. "Now that is a scientific proof!"

Their friendship was never harmed by these discussions...

One day "the Woman" arrived in the form of a widow who was nearing 40 years old. The lady had a few pieces of land that basked in the sun in summer and froze in winter. The arms and the pension of the free thinker were not without their useful charm. The couple was brought together by third parties and a marriage was agreed to. The bonesetter approved of it as soon as he learned he would be a welcome guest in the new household.

Reason has always struggled to master feelings. The free thinker thought about his long celibacy and began to doubt, not his desire for his fiancée but his ability to express it. As often happens, doubt fosters doubt and doubt leads to an inability to act.

The free thinker told the bonesetter of his fear of being made impotent by doubt.

The evening before the wedding, a night made crystal clear by the January mistral, the bonesetter invited his friend to a final meal between bachelors. For the occasion, he prepared scrambled eggs with truffles. The dish was a little rough, but it was strongly flavored with three magnificent truffles, which were grated into the eggs with a great deal of affection.

The free thinker was touched by this thoughtfulness. He ate the entire omelet by himself, for his friend would not, of course, touch it.

The next day he wed the widow.

Ten moons later the first triplets in the region were born in the house of the free thinker, and they were magnificent.

The bonesetter went to congratulate his friend and insisted: "It's scientific, I tell you!"

The free thinker wondered until his dying day what he would have had if the omelet had been made with five truffles!

Index of Recipes

Fish

The Truffle on Its Own

Stuffed Things

Poultry

Meats

Sauces

Vegetables

Rice and Pasta

Cheese

Desserts